JOURNAL OF GREEN ENGINEERING

Volume 5, No. 1 (October 2014)

Special Issue on:

Green Mobile Communication

Guest Editor:
Prithiviraj Venkatapathy

JOURNAL OF GREEN ENGINEERING

Chairperson: Ramjee Prasad, CTIF, Aalborg University, Denmark
Editor-in-Chief: Michele Albano, ISEP - Instituto Superior de Engenharia do Porto, Portugal

Editorial Board
Luis Kun, National Security, National Defense University/Center for Hemispheric Defense Studies, USA
Dragan Boscovic, Motorola, USA
Panagiotis Demstichas, University of Piraeus, Greece
Afonso Ferreira, CNRS, France
Meir Goldman, Pi-Sheva Technology & Machines Ltd., Israel
Laurent Herault, CEA-LETI, MINATEC, France
Milan Dado, University of Zilina, Slovak Republic
Demetres Kouvatsos, University of Bradford, United Kingdom
Soulla Louca, University of Nicosia, Cyprus
Shingo Ohmori, CTIF-Japan, Japan
Doina Banciu, National Institute for Research and Development in Informatics, Romania
Hrvoje Domitrovic, University of Zagreb, Croatia
Reinhard Pfliegl, Austria Tech-Federal Agency for Technological Measures Ltd., Austria
Fernando Jose da Silva Velez, Universidade da Beira Interior, Portugal
Michel Israel, Medical University, Bulgaria
Sandro Rambaldi, Universita di Bologna, Italy
Debasis Bandyopadhyay, TCS, India

Aims and Scopes
Journal of Green Engineering will publish original, high quality, peer-reviewed research papers and review articles dealing with environmentally safe engineering including their systems. Paper submission is solicited on:

- Theoretical and numerical modeling of environmentally safe electrical engineering devices and systems.
- Simulation of performance of innovative energy supply systems including renewable energy systems, as well as energy harvesting systems.
- Modeling and optimization of human environmentally conscientiousness environment (especially related to electromagnetics and acoustics).
- Modeling and optimization of applications of engineering sciences and technology to medicine and biology.
- Advances in modeling including optimization, product modeling, fault detection and diagnostics, inverse models.
- Advances in software and systems interoperability, validation and calibration techniques. Simulation tools for sustainable environment (especially electromagnetic, and acoustic).
- Experiences on teaching environmentally safe engineering (including applications of engineering sciences and technology to medicine and biology).

All these topics may be addressed from a global scale to a microscopic scale, and for different phases during the life cycle.

Published, sold and distributed by:
River Publishers
Niels Jernes Vej 10
9220 Aalborg Ø
Denmark

River Publishers
Lange Geer 44
2611 PW Delft
The Netherlands

Tel.: +45369953197
www.riverpublishers.com

Journal of Green Engineering is published four times a year.
Publication programme, 2015: Volume 5 (4 issues)

ISSN 1904-4720 (Print Version)
ISSN 2245-4586 (Online Version)
ISBN 978-87-93237-96-4

JOURNAL OF GREEN ENGINEERING

Volume 5 No. 1 October 2014

Editorial Foreword:
Special Issue on Green Mobile Communication

A major challenge for Information and Communication (ICT) has been the reduction of power consumption in telecommunication networks. In the present day scenario the energy demand from major mobile networks is of the order of several thousand Giga watts per hour per year. The escalating energy prices has led to a situation where energy expenses equal around 18% of the network operational cost in European markets and it much more in developing countries like India where diesel fuel has been utilized to power radio base stations for off grid power operations. This has resulted for many service providers in India to rely on backup generators to ensure that network is up and running and the resultant expenditure amounts to one third of their budget for diesel fuel utilization. It is estimated that ICT is accountable for 2 to 4 percent of the worldwide carbon emissions. The power consumption during the usage phase of equipment accounts for roughly 40 to 60 percent of the carbon emissions. By 2020 these emissions are expected to double if no clear initiatives are taken to reduce this footprint.

Lastly, ICT is being regarded as a solution with the potential to eliminate 15 percent of the global carbon footprint. In India, currently there are nearly 1 billion cell phone users and nearly 10 lakh cell phone towers (BTS) to meet the communication demands and the carbon footprint from these mobile towers within India is estimated to be approximately 13 million tonnes.

There are 3 strong motivating factors that drive further research and development in reducing the power consumption of cellular networks:

A. The first factor which is crucial is to minimize the environmental impact on this sector for climate change caused by increased CO_2 and other Green house gases concentration level in the atmosphere which is caused due to air conditioning equipment or the use of fossil fuels for producing electrical energy.

B. The second motivating factor is due to the high power Electromagnetic Radiation (EMR) directed from the cellular towers which results in health hazards for human beings and other living species including flora and fauna. It is a general opinion that EM radiation is harmful to mankind which could cause neurological, respiratory, cardiac, ophthalmological and other conditions ranging in severity from headaches, fatigue and add to psychosis and strokes.

C. Besides their corporate responsibility regarding environmental protection, cellular network providers are also becoming aware of their energy bills which can translate from 18% (EU) to 32% (India) for their operational expenses (OPEX).

Thus reduction of energy consumption in cellular networks will translate into direct economic benefits like cost reduction. This trend has stimulated the interest of researchers into an innovative new research area designated as "Green Cellular Networks".

Green Communication

Green communication technique intends to reduce the power consumption of the cellular networks and the electromagnetic pollution due to them. Green communication is vast research discipline that needs to cover all the layers of the protocol stack and it is important to identify the fundamental tradeoffs linked with energy efficiency and the overall performance of the network.

In this special issue on Green Mobile Communication the following four papers are featured:

- Quad band Signal Strength Monitoring System Using Quadcopter and Quadphone
- Setting up a low cost sustainable telecommunication infrastructure for rural Communications in developing countries
- Cross Layer Design Based Green Cellular Architecture Using Stochastic Optimization
- Scheduling BTS Power Levels For Green Mobile Computing

Quad band Signal Strength Monitoring System Using Quadcopter and Quadphone

The communication protocols used in Base transceiver station (BTS) could be harmful to human species and other life forms in the ecosystem. The BTS used for these systems could emit radiations beyond safety threshold. Therefore, it is essential to monitor such power radiation level from time to time. Manual readings at each BTS are strenuous work and time consuming. This paper proposes radiation measurement using quadphone mounted within the quadcopter working on android software platform. Quadcopter is a low cost restricted payload machine which suits the measurement of radiation emitted from antenna towers and power lines. An application is developed to monitor the power radiations emitted by each of the cellular mobile bands and the associated communication protocol by the quadphone. A Quadphone incorporating such an application is used here to record power radiation levels at several points around a single BTS tower. The combined operation of the Quadphone and Quadcopter can be used effortlessly to record power readings for each BTS in the shortest duration. A signal strength monitoring system is developed and field test was carried out to measure signal strength in dbm, SNR ratio and EVDO value in this paper.

Cross Layer Design based Green Cellular Architecture Using Stochastic Optimization

Cross layer techniques are in general used to enhance a network's perfor-mance. Various cross layer models have been proposed by researchers for energy efficient scenarios, but most of these models do not consider all the fundamental Quality of service requirements along with energy efficiency. Quality of Service and Queue Stability affect the energy consumption and network performance in each time slot of a network. So an adaptive model is necessary to guarantee the Quality of service and Queue stability along with reduced energy consumption. The model proposed in this paper uses the stochastic drift plus penalty method to improve energy efficiency along with Quality of Service and Queue stability constraints. The optimization technique in the proposed model does not require channel density function. The energy efficiency improvement under Quality of Service and Queue Stability constraint is demonstrated by simulation studies in the paper.

Setting up a Low Cost Sustainable Telecommunication Infrastructure for Rural Communications in Developing Countries

This paper provides information on Government of India initiative to provide Broadband services to Rural India. In the absence of reliable grid supply in rural areas, deployment of alternate power system plays major role to power the Broadband systems deployed in rural areas. Service expansion requires fast site setup and low power consumption. This is being challenged by site acquisition and construction issues because of inconvenient access to many rural areas and energy supply that is often unstable or cannot keep up with demand. This contribution paper provides methodology based on the experience of the actual on-site techniques adopted in Broadband Network resulting in huge savings on operating expenses due to fuel consumption of diesel gen-sets and to optimize & harvest maximum renewable energy from available renewable sources in the rural areas having either no grid or poor grid.

Scheduling BTS Power Levels for Green Mobile Computing

Considering the objectives of Green Mobile Computing, this work proposes to dynamically allocate BTS's transmission power level according to the requirement of number of users in a particular radio cell. The number of users is more during the daytime when compared to night time, so the power level can be dynamically reduced to overcome the hazardous effects of radiation. A scheduling algorithm for switching the transmission power level at BTS, based on the output derived from neural network, which is trained with historical data collected from local authorities to learn the population pattern and assign corresponding power levels. The scheduling algorithm with Artificial Neural Network (ANN) gives reduced power consumption, low interference and shortened radiation exposure.

Prof. Dr. V. Prithiviraj Principal,
RIT, Chennai, India

Quad Band Signal Strength Monitoring System Using Quadcopter and Quad Phone

Prem Kumar N[1], Raj Kumar A[1], Sundra Anand[1],
Dr. E. N. Ganesh[2] and Dr. V. Prithiviraj[3]

[1]*Final year B.E. (ECE), Rajalakshmi Institute of Technology, Tamil Nadu, India*
[2]*Dean R&I, Rajalakshmi Institute of Technology, Tamil Nadu, India*
[3]*Professor, Dept. of Electronics and Communication Engg (ECE),
Rajalakshmi Institute of Technology, Tamil Nadu, India
Corresponding Authors: {prem0504kumar; raj16993; sundra.ece;
profvpraj}@gmail.com; dean.research@ritchennai.ediu.in*

Received 2 Feb 2015; Accepted 5 March 2015;
Publication 29 May 2015

Abstract

The communication protocols used in Base Transceiver Station (BTS) could be harmful to human species and other life forms in the ecosystem. The BTS used for these systems could emit radiations beyond safety threshold. Therefore, it is essential to monitor such power radiation levels from time to time. Manual readings at each BTS are strenuous and time consuming work. This paper proposes radiation measurement using Quad Phone assembled within the Quadcopter using android software. Quadcopter is a low cost restricted pay-load machine which suits the measurement of radiation emitted from antenna towers and power lines. An application is developed to monitor the power radiation emitted by each of the bands and the associated communication protocol by utilizing the Quad Phone. It can use CDMA, GSM, 3G and LTE protocols at the designated frequency bands. A Quad Phone incorporating the android application is used to record power radiation levels at several points around a single tower facilitated by the Quadcopter flight navigation system. The Wave Point navigation is a hardware module used to move around a target point up to 16 locations and circle back to the initial position. The collected

Journal of Green Engineering, Vol. 5, 1–22.
doi: 10.13052/jge1904-4720.511

data regarding the radiated signal strength for different protocols are either transmitted through wireless or stored within the Quad Phone which could be retrieved later. The power readings around each base station can be recorded using the Quad Phone and the Quadcopter in the shortest time possible. Hence the combination of the Quad Phone mounted on the Quadcopter provides an excellent monitoring system for auditing the Electromagnetic Radiation(ER) and subsequently determine the Electromagnetic Pollution Index (EPI) from the delineated pockets of pollution regions.

Keywords: Base Transceiver Station (BTS), Code Division Multiple Access (CDMA), Global system for Mobile Communications (GSM), Wave point navigation, Global Positioning System (GPS), Electromagnetic Radiation(ER).

1 Introduction

Quadcopter is a multirotor helicopter and is propelled by four rotors. Quadcopters do not require mechanical linkages to vary the rotor blade pitch angle as they spin. This simplifies the design and maintenance of the vehicle. The use of four rotors allows each individual rotor to have a smaller diameter than the equivalent helicopter rotor, allowing them to possess less kinetic energy during flight. There is currently a limited diversity of vehicles capable of Vertical Take-Off and Landing [1]. The Base Transceiver Stations emits radiations which is being picked up by the cell phones and enables to establish a connection. Now-a-days in order to increase the cell coverage the service providers could radiate large amount of energy which could be harmful to the human species and other flora and fauna. To monitor over these radiation levels the Indian Ministry of Telecommunication formed a committee called TERM cells where the members of the committee move around each BTS measuring the signal strength for each service providers for every quarter. This method is found to be tedious as it requires large amount of human resource, vehicle field strength measurement and is a time consuming process.

This brought immediate attention and as a solution for this problem it is proposed to develop a frame work with a goal to develop an android application for monitoring the signal strength of the BTS at various locations utilizing a Quad Phone mounted within a Quadcopter. A quad phone is quad core quad band phone that works in multiple radio frequency bands using GSM/CDMA technology. The need for Quadphone is to measure the signal strength across four different cellular services like GSM, CDMA, HSDPA and LTE. Since Quad Phone is light weight smart phone it can be readily used with a Quadcopter to measure the signal strength and radiation level. A Quadcopter is an aircraft that becomes airborne due to the lift force provided by four rotors usually mounted in cross configuration. It is an entirely different vehicle when compared with a helicopter, mainly due to the way both are controlled. Quadcopters are unmanned air vehicles which are classified into two main groups which are heavier-than-air and lighter-than-air [1]. These two groups self-divide in many other that classify aircrafts according to motorization, type of lift off and many other parameters. Vertical Take-Off and Landing (VTOL) UAVs like Quadcopters have several advantages over fixed-wing airplanes. They can move in any direction and are capable of hovering and fly at low speeds. Vertical takeoff landing can be implemented in Quadcopter with efficient design. Given these characteristics, Quadcopters can be used in search and rescue missions, meteorology, penetration of hazardous environments (e.g. exploration of other planets) and other applications suited for such an aircraft. Quadcopter finds extensive applications in research areas like control engineering, security, remote sensing and disaster management scenario etc.

Each rotor in a Quadcopter is responsible for a certain amount of thrust and torque about its center of rotation, as well as for a drag force opposite to the rotorcraft's direction of flight. The Quadcopter's propellers are not all alike. In fact, they are divided in two pairs, two pushers and two puller blades that work in contra-rotation. As a consequence, the resulting net torque can be null if all propellers turn with the same angular velocity, thus allowing for the aircraft to remain still around its center of gravity. In order to define an aircraft's orientation (or attitude) around its center of mass, aerospace engineers usually define three dynamic parameters, the angles of yaw, pitch and roll. This is very useful because the forces used to control the aircraft act around its center of mass, causing it to pitch, roll or yaw.

Figures 1 and 2 illustrates the movements of each rotor. Changes in the pitch angle are induced by contrary variation of speeds in propellers 1 and 3 (see Figure 1), resulting in forward or backwards translation. If we do this same action for propellers 2 and 4, we can produce a change in the roll angle and we will get lateral translation. Yaw is induced by mismatching the balance in aerodynamic torques (i.e. by offsetting the cumulative thrust between the counter-rotating blade pairs). So, by changing these three angles in a Quadcopter we are able to make it maneuver in any direction.

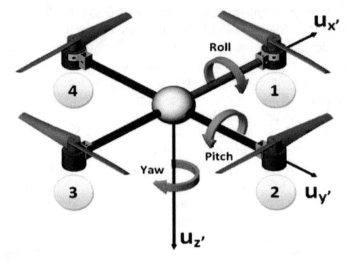

Figure 1   Yaw, pitch and roll rotations of a Quadcopter.

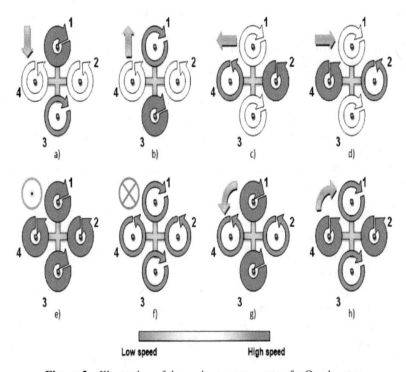

Figure 2   Illustration of the various movements of a Quadcopter.

2 Quadcopter Design

The most important target of this particular design process is to arrive at the correct set of requirements for the copter with appropriate payload which could be summarized into a set of specifcations. The specifcations for the Quadcopter prototype are given below:

- Flight autonomy between 15 to 25 minutes.
- On-board controller should have its separate power supply to prevent simultaneous engine and processor failure in case of battery loss.
- Ability to transmit live telemetry data and receive movement orders from a ground station wirelessly, therefore avoiding the use of cables which could become entangled in the aircraft and cause an accident;
- Quadcopter should not fly very far from ground station so there is no need of long range telemetry hardware and the extra power requirements associated with long range transmissions.

Consequently, the main components are:

- 4 electric motors and 4 respective Electronic Speed Controllers (ESC).
- 4 propellers.
- 1 on-board processing unit with inbuilt Accelerometer, Gyroscope, Barometer and Magnetometer.
- 2 on-board power supplies (batteries), one for the motors and another to the processing unit. At this point we will assume that beyond the need to provide electrical power to the motors, we must assure that the brain of the Quadcopter (i.e. the on-board processing unit) remains working well after the battery of the motors has discharged;
- 1 airframe for supporting all the aircraft's components.

2.1 Airframe

The airframe is the mechanical structure of an aircraft that supports all the components, much like a "skeleton" in Human Beings. Designing an airframe from scratch involves important concepts of physics, aerodynamics, materials engineering and manufacturing techniques to achieve certain performance, re- liability and cost criteria. We have designed the airframe of our Quadcopter with 258g of mass and made of GPR (Glass-Reinforced-Plastic), possessing a cage-like structure in its center that will offer extra protection to the electronics.

This particular detail may prove itself very useful when it comes to the test flight stage, when accidents are more likely to happen.

Figure 3   Top and the bottom plate.

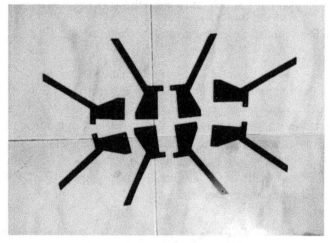

Figure 4   Quadcopter legs.

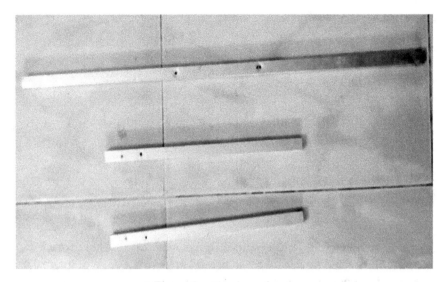

Figure 5   Aluminum frame.

Figure 6   Quad-X frame.

2.2 Propellers

The typical behavior of a propeller can be defined by three parameters:

- Thrust coefficient c_T;
- Power coefficient c_P;
- Propeller radius r.

These parameters allow the calculation of a propeller's thrust T:

$$T = \frac{c_T 4\rho r^4 \omega^2}{\pi^2} \tag{1}$$

and power

$$P_P{:} = \frac{c_P 4\rho r^5 \omega^3}{\pi^3} \tag{2}$$

where ω is the propeller angular speed and ρ the density of air. These formulas show that both thrust and power increase greatly with propeller's diameter. If the diameter is big enough, then it should be possible to get sufficient thrust while demanding low rotational speed of the propeller. Consequently, the motor driving the propeller will have lower power consumption, giving the Quadcopter higher flight autonomy. Available models of counter rotating propellers are scarce in the market of radio controlled aircrafts.

The "EPP1045" (see Figure 7), a propeller with a diameter of 10″ (25.4 cm) and weighting 23g, presented itself as a possible candidate for implementation in the Quadcopter. To check its compatibility with the project requisites it is necessary to calculate the respective thrust and power co-efficient. We can extract the mean thrust and power co-efficient by using Equations (1) and (2):

$$c_T = 0.1154 \tag{3}$$

$$c_P = 0.0743 \tag{4}$$

In reality, neither the thrust nor the power co-efficient are constant values, they are both functions of the advance ratio J:

$$J = u_0 n D_P \tag{5}$$

where u_0 is the aircraft flight velocity, n the propeller's speed in revolutions per second and finally D_P is the propeller diameter.

However, when observing the characteristic curves for both these co-efficient as shown in Figures 8 and 9, it is clear that when an aircraft's flight

Figure 7 EPP1045 propellers.

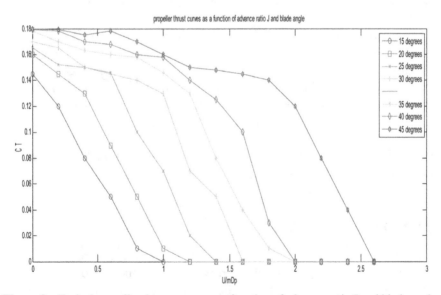

Figure 8 Typical propeller thrust curves as a function of advance ratio J and blade angle.

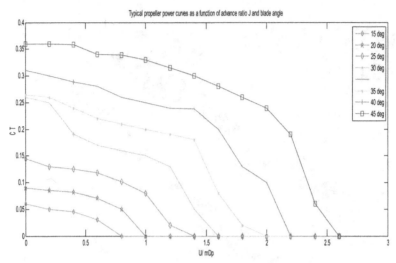

Figure 9   Typical propeller power curves as a function of advance ratio J and blade angle.

velocity is very low (e.g. in a constant altitude hover) the advance ratio is almost zero and the two co-efficient can be approximated as constants, which is the current case, for at this point there is no interest in achieving high translation velocity. Assuming the Quadcopter's maximum weight is 9.81N (1kg) and that we have four propellers, it is mandatory that each propeller is able to provide at least 2.45N (1/4 of the Quadcopter weight) in order to achieve lift-off. Taking this data into consideration leads us to wonder about the minimum propeller rotational speed involved, as well as the magnitude of the power required for flight. Figure 10 helps us with some of these questions. It is observed that a propeller will have to achieve approximately 412rad.s^{-1}, which is equivalent to 3934 revolutions per minute, to provide the minimum 2.45N required for lift-off [6].

The respective propeller power is 26W. After this short analysis we can state that the EPP1045 propellers are suitable for implementation in the Quadcopter prototype.

3 Multi-Wii Configuration

The Multi Wii Copter is an open source software which controls the RC Platform and also compatible with the hardware boards and sensors.

The first and most famous setup is the association of a Wii Motion Plus and a Arduino pro mini board. The MultiWii 328P is a gyro/accelerometer

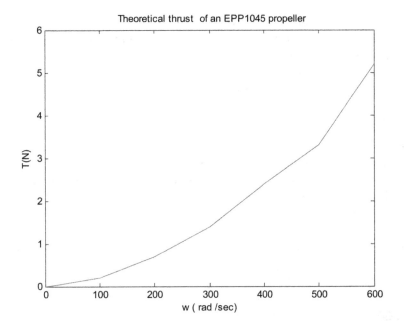

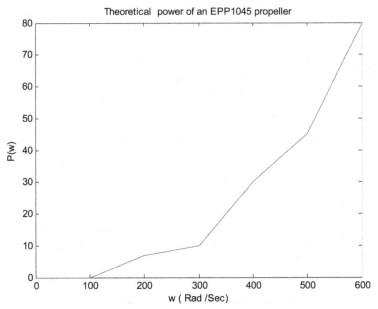

Figure 10   Theoretical thrust and power of an EPP1045 propeller.

based flight controller that is loaded with features. This version of the MultiWii supports DSM2 satellite receiver functionality. With expandability options and full programmability, this device can control just about any type of aircraft. This is the ideal flight controller for your multi-rotor aircraft. The pin diagram for MultiWii 328P is given in Figure 11.

3.1 A Gimbals

A gimbals is a pivoted support that allows the rotation of an object about a single axis. A set of three gimbals, one mounted on the other with orthogonal pivot axes, may be used to allow an object mounted on the innermost gimbal to remain independent of the rotation of its support. For example, on a ship, the gyroscopes, shipboard compasses, stoves, and even drink holders typically use gimbals to keep them upright with respect to the horizon despite the ship's pitching and rolling. When associated with an accelerometer, MultiWii is able to drive 2 servos for PITCH and ROLL gimbal system adjustment. The gimbal can also be adjusted via 2 RC channels.

3.2 B ITG3205 Triple Axis Gyro

This is a breakout board for InvenSense's ITG-3205, a groundbreaking triple-axis, digital output gyroscope. The ITG-3205 features three 16-bit analog-to-digital converters (ADCs) for digitizing the gyro outputs, a user-selectable internal low-pass filter bandwidth, and a Fast-Mode I2C (400kHz) interface. Additional features include an embedded temperature sensor and a 2% accurate internal oscillator. The ITG-3205 can be powered at anywhere

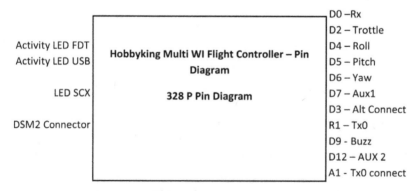

AO – CAM Pitch A1 – CAM Roll A2 –CAM Trigger   D3, D10 – Motor Connect

Figure 11   Pin diagram of MultiWii 328P.

between 2.1 and 3.6V. For power supply flexibility, the ITG-3205 has a separate VLOGIC reference pin (labeled VIO), in addition to its analog supply pin (VDD) which sets the logic levels of its serial interface. The VLOGIC voltage may be anywhere from 1.71V min to VDD max. For general use, VLOGIC can be tied to VCC. The normal operating current of the sensor is just 6.5mA.

Communication with the ITG-3205 is achieved over a two-wire (I2C) interface. The sensor also features a interrupt output, and an optional clock input.

A jumper on the top of the board allows you to easily select the I2C address, by pulling the AD_0 pin to either VCC or GND; If CLKIN pin is not used jumper shoul be shorted on the bottom of the board to tie it to GND.

3.3 C BMA180 Accelerometer

This is a breakout board for Bosch's BMA180 three-axis, ultra-high performance digital accelerometer. The BMA180 provides a digital 14-bit output signal via a 4-wire SPI or I2C interface. The full-scale measurement range can be set to ±1g, 1.5g, 2g, 3g, 4g, 8g or 16g. Other features include programmable wake-up, low-g and high-g detection, tap sensing, slope detection, and self-test capability. The sensor also has two operating modes: low-noise and low-power.

This breadboard friendly board breaks out every pin of the BMA180 to an 8-pin, 0.1″ pitch header. The board doesn't have any on-board regulation, so the provided voltage should be between 1.62 and 3.6V for VDD and 1.2 to 3.6V for VDDIO. The sensor will typically only consume 650uA in standard mode.

3.4 D BMP085 Barometer

This precision sensor from Bosch is the best low-cost sensing solution for measuring barometric pressure and temperature. Because pressure changes with altitude you can also use it as an altimeter! The sensor is soldered onto a PCB with a 3.3V regulator, I2C level shifter and pull-up resistors on the I2C pins.

3.5 E HMC5883L Magnetometer

The Honeywell HMC5883L is a surface-mount, multi-chip module designed for low-field magnetic sensing with a digital interface for applications such as low-cost compassing and magnetometer. The HMC5883L includes our state-of-the-art, high-resolution HMC118X series magneto-resistive sensors plus an ASIC containing amplification, automatic degaussing strap drivers, offset cancellation, and a 12-bit ADC that enables 1° to 2° compass heading accuracy. The I2C serial bus allows for easy interface. The HMC5883L is a $3.0 \times 3.0 \times 0.9$ mm surface mount 16-pin leadless chip carrier (LCC). Applications for the HMC5883L include Mobile Phones, Netbooks, Consumer Electronics, Auto Navigation Systems, and Personal Navigation Devices.

The HMC5883L utilizes Honeywell's Anisotropic Magneto Resistive (AMR) technology that provides advantages over other magnetic sensor technologies. These anisotropic, directional sensors feature precision in-axis sensitivity and linearity.

4 Graphical User Interface

Java language is used to code in Linux platform. GUI is developed for graphical visualization of sensors, processing control of motors in Quadcopter utilizing RC Signalling.

Figure 12 shows the Multi Wii Simulation for Quadcopter with GUI. It gives flying path, speed, latitude and longitude

5 Android Application (SSM)

Android application is developed for measuring the signal strength through Quad Phone. Android app created in the phone has user interface with the following parameters are created.

- IMEI Number
- Cell ID
- Signal Strength in dBm
- EVDO Value
- SNR Value
- Button to fetch Details

Once this button is clicked all the details are displayed on the screen. Android app requires minimum of 1 Mb memory requirement. All these contents are visible in the User Interface of the SSM application on the quad phone, which is placed over the Quadcopter as shown in Figures 13 and 14.

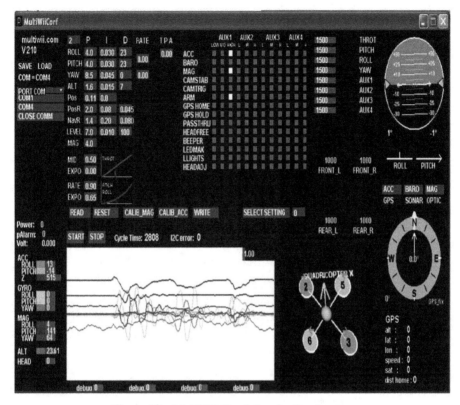

Figure 12 MultiWii Simulation for Quadcopter.

In order to view the signal strength displayed on the screen of the quad phone in PC, a new application BBQ is used. This BBQ requires the client software to be installed in the PC which is used to view the screen of the mobile phone [3]. BBQ is installed with the demo shown in Figure 13. Now the Quad Phone on the Quadcopter and the PC at the service end are connected to Wi-fi.

When BBQ is started it connects to IP automatically and this IP is connected to Mobile phone. Prior to this IP is given in BBQ software as shown in Figure 14 and thus a wireless connection between the PC and the quad phone is created. Using this wireless connection the screen of the quad phone can be viewed on the monitor of PC. Figure 15 shows the final picture of Quadcopter with Quad Phone. Figures 16(a) and 16(b) are screen shot of BBQ for measuring the signal strength.

Figure 13   Mobile version of the BBQ application.

6 Results and Conclusion

In this project, the development of the hardware and software framework necessary is undertaken to enable the Quadcopter to fly autonomously. The associated mechanical and electrical hardware is assembled and tested for its viability. The efficient design of Quadcopter housing the Quad Phone is utilized to measure the Signal Strength, SNR and and also to display the IMEI, cell ID and EVDO values associated with the Quad Phone. On analysis it is found that the proposed method of Quadcopter design developed provides a document that clearly and precisely outlines the steps necessary to assemble and fly the Quadcopter. With the implemented control scheme, the Quadcopter is able to hover autonomously and perform step movements in all directions. The experiments for several testing flying session have been

BBQScreen - Connect

BBQScreen

Welcome to BBQScreen client app! This apps let you see your screen shared by the BBQScreen Server App running on your Android device.

To get started: Turn on the 'BBQScreen Remote Control' app on your Android device. It should appear in the list aside shortly after.

Make sure: That you are on the same Wi-Fi network as your device, that your device screen is unlocked when connecting, and that the device IP address (if entered manually) is correct.

List of detected devices

Or manually enter IP: 192.168.1 .101

CONNECT

Client settings and information

☑ HD processing (slower)

☐ Show FPS

Bootstrap non-root USB method

Go to our website for more information about USB method

Shortcuts : CTRL + 'O' : Offset orientation by -90°
'P' : Offset orientation by 90°
'F' : Full-screen toggle

Go to our website

Client version 2.2.2

Figure 14   PC version of the BBQ client to view the phone screen on PC.

Figure 15   Final picture of the proposed Quadcopter.

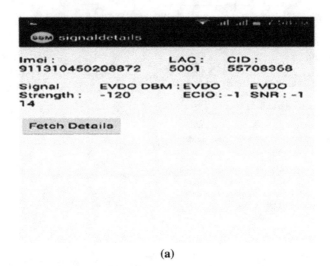

(a)

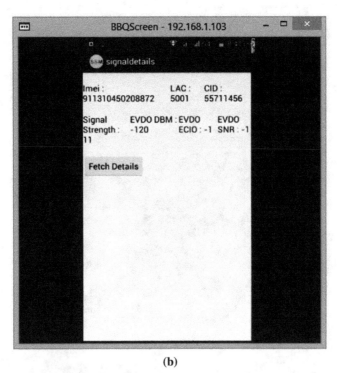

(b)

Figure 16 (a) Screenshots of the SSM Application over cell phone, (b) Screenshot of the application shared over the laptop.

performed for tuning the weight matrices of the controller and to carry out performance tests.

The main objective of this paper is to measure the radiated signal strength from a single Base Transceiver Station. This is achieved by mounting Quad Phone on a Quadcopter. An android application named "Signal Strength Monitoring (SSM)" is developed to monitor the power radiations emitted by each band i.e CDMA, GSM, HSPA, LTE in the communication protocol. So the mobile phone incorporating such an application is mounted on a Quadcopter to record the power radiation levels at several points around a single tower. Two signal strength values of 14 dbm and 11 dbm is noted for two different field tests carried out with SNR of -1 and EVDO value of -120. The Signal Strength that is displayed by the Android Application (SSM) is successfully shared over a laptop using BBQ Software using the Wi-Fi connection and IP address.

The above design encompassing a Quad Phone within the Quadcopter could pave way for undertaking Electromagnetic Pollution Index surveys within sensitive zones, City Malls, Railway Stations, Hospital zones and Airport restricted areas etc., [5]. This technique would carve out a distinct possibility for monitoring the Electromagnetic radiation in a residential complex like multi storied buildings/flats which are directly in the line of sight of the radiating tower antennas catering to various service providers. The combination of the Quadcopter and the Quad Phone would enable auditing of the Electromagnetic Radiation(ER) and subsequently determine the Electromagnetic Pollution Index (EPI) from the delineated pockets of pollution regions.

References

[1] Claudia Mary; Luminita Cristiana Totu; SimonKonge Koldbæk (June 2010), "Modelling and Control of Autonomous Quad-Rotor", Aalborg University.

[2] Jeremia, S.; Kuantanna, E.; Pangaribuan, J., (Sept. 2012), "Design and construction of remote-controlled quad-copter based on STC12C-5624AD", System Engineering and Technology (ICSET), 2012 International Conference, page(s):1–6.

[3] Malathi, S.; TirumalaRao, G.; Rajeswer Rao, G., (December 2013), "A Prediction Model for Electromagnetic Pollution Index of Multi System Base Stations", Internation Journal of Engineering Research and Technology (IJERT), Vol. 2 Issue 12.

[4] Mattar, N.A.B.; Razak, M.R.B.A.; Murat, Z.B.H.; Khadri, N.B.; Rani, H.N.B.H.M., (2002), "Measuring and analyzing the signal strength for Celcom GSM [019] and Maxis [012] in UiTM Shah Alam campus", Research and Development, 2002. SCOReD 2002, page(s): 489–493.

[5] Dr. Prithiviraj Venkatapathy; Cmde J Jena. N; Capt. Avadhanulu Jandhyala, (April 2012), "Electromagnetic Pollution Index- A Key Attribute of Green Mobile Communications", Green Technologies Conference, 2012 IEEE, Page(s): 1–4.

[6] Zhang Yao; Tianjin Key Lab. of Process Meas. & Control, Tianjin Univ., Tianjin, China; Xian Bin; Yin Qiang; Liu Yang, (2012), "Autonomous control system for the quadrotor unmanned aerial vehicle", Control Conference (CCC), 2012 31st Chinese, Page(s): 4862–4867.

Biographies

N. P. Kumar obtained his Bachelor of Engineering in Electronics and Communication Engineering from Rajalakshmi Institute of Technology, Chennai in 2014. Currently he is working as Backup Administrator. His areas of interests are Electronics & Circuits, Digital Circuitry and Robotics.

A. R. Kumar has completed his Bachelor of Engineering (Electronics and Communication Engineering) in 2014 from Rajalakshmi Institute of Technology, Chennai. Currently he is working as Assistant System Engineer in

Tata Consultancy Services. His areas of interest include Robotics, Digital Electronics and Mobile Communication.

S. Anand obtained his Bachelor of Engineering in Electronics and Communication Engineering from Rajalakshmi Institute of Technology, Chennai in 2014. Currently he is working as Engineer-Trainee. His areas of interests include Networking and Digital Circuitry.

E. N. Ganesh received M.Tech degree in Electrical Engineering from IIT Madras, Ph.D. from JNTU Hyderbad. He has over 20 years of academic experience and now working as Dean (Research and Innovation) at Rajalakshmi Institute of Technology. His area of interests is Nanoelectronics, Robotics and Hyperspectral Image Processing.

V. Prithiviraj received M.S degree in Electrical Engineering from IIT Madras., Ph.D. in Electronics and Electrical Communication Engineering from IIT Kharagpur. He is working as Principal Rajalakshmi Institute of Technology from May 2013. He has over 3 decades of teaching experience and 12 years of Research & Development Experience between the two IITs in the field of RF & Microwave Engineering. His areas of interest include Broadband and Wireless Communication, Telemedicine, e-Governance and Internet of Things.

Cross Layer Design based Green Cellular Architecture Using Stochastic Optimization

L. Senthil Kumar[1], J. Vasantha Kumar[2]
and M. Meenakshi[3]

[1]Research Scholar, CEG, Anna University, Tamil Nadu, India
[2]Post-Graduate Student, CEG, Anna University, Tamil Nadu, India
[3]Professor, CEG, Anna University, Tamil Nadu, India
Corresponding Authors: senthilkumarl@live.com; j.vasanth489@gmail.com;
meena68@annauniv.edu

Received 2 Feb 2015; Accepted 5 March 2015;
Publication 29 May 2015

Abstract

Cross layer techniques are in general used to enhance a network's per-formance. Various cross layer models have been proposed by researchers for energy efficient scenarios, but most of these models do not consider all the fundamental Quality of service requirements along with energy efficiency. Quality of Service and Queue Stability affect the energy con-sumption and network performance in each time slot of a network. So an adaptive model is necessary to guarantee the Quality of service and Queue stability along with reduced energy consumption. The model proposed in this paper uses the stochastic drift plus penalty method to improve energy efficiency along with Quality of Service and Queue stability constraints. The optimization technique in the proposed model does not require chan-nel density function. The energy efficiency improvement under Quality of Service and Queue Stability constraint is demonstrated by simulation studies in the paper.

Journal of Green Engineering, Vol. 5, 23–48.
doi: 10.13052/jge1904-4720.512

Keywords: Cross layer, Quality of Service, stochastic drift plus penalty method, Energy efficiency, Queue Stability.

1 Introduction

Cross layer design based network performance improvement has been an evolving strategy in recent times. Though many adaptation schemes are deployed in different OSI layers, the lack of coordination among them makes the overall performance of the system non-optimal. Only proper coordination across layers can benefit the system to achieve Quality of Service (QoS) with optimized goals across layers. For some applications, the packet arrival rate at the transmission buffer is continuous, while for some other applications, the packet arrival is quite bursty in nature. Therefore, if the packet scheduler at the lower layer does not utilize the traffic information of the application it is dealing with, it may cause excessive delay (and buffer overflow when the buffering capacity is limited) and/or excessive power consumption. An intelligent packet scheduler should be able to adjust the transmission rate at the physical layer depending not only on the channel gain, but also on the buffer while satisfying the QoS requirements on delay, overflow and packet error rate. For example, when packet delay is relatively less important than transmission power, the scheduler should not hurry up transmission by using a higher power level in bad channel conditions when the buffer has relatively fewer packets. It can wait for a better channel condition.

This approach achieves two goals: it satisfies the packet error rate, delay and buffer overflow requirements and it does so with the lowest possible transmission power. In future green radio networks, the scheduler will have to apply similar techniques to save energy. In this paper, we show how joint optimization can be used in an intelligent scheduler to reduce energy consumption.

Among all the cross-layer adaptation techniques, the rate and power adaptation techniques at the physical layer are the most important ones for green radio network design, since they minimize transmission power based on upper layer information. Therefore, without loss of generality, in this paper, we concentrate on the power minimization issue that is of particular importance for green radio networks. We show how the transmitter power can be saved using cross-layer optimal policies, where the rate and power at the physical layer are adjusted to minimize power with specific QoS requirements, thereby striking a balance between "green needs" and service requirements.

The objective of this work is to design a cross-layer scheduler that determines the number of packets to be transmitted in each time slot. In order to determine number of packets, the scheduler utilizes both the physical layer information (e.g., channel gain) and the data-link layer information (e.g., buffer occupancy and nature of traffic). Also, scheduler objectives are to minimize average transmission power under the constraints on the average delay and average overflow. Therefore, in a particular time-slot n, the scheduler first determines the states of the traffic, buffer and channel, and then chooses action dynamically to optimize average transmission power, delay and overflow. Since the nature of the problem is dynamic, it falls into the general category of *stochastic dynamic programming* problems. To solve this dynamic problem, we use Lyaponuv drift plus penalty algorithm where in the number of packets to be transmitted in each time slot is determined for transmission over fading channels considering both the physical layer and the data-link layer optimization goals. At the physical layer, our goal is to optimize the transmission power while satisfying a particular bit error rate (BER) requirement. On the other hand, at the data-link layer, our goal is to optimize the delay and packet loss due to overflow. Overall, the cross-layer approach is shown to be effective in conserving the energy of the system while satisfying the QoS requirements.

1.1 Related Works

Energy-efficient cross-layer optimized techniques and designs have been a major research attention in the last decade among wireless researchers working in different networks and protocol stacks.

In [1], the authors have presented a cross-layer design technique that determine the optimal policy based on both the physical layer and the data-link layer information with cross-layer dynamic adaptation policy, thereby delay, overflow rate and BER also be guaranteed precisely for all traffic arrival rates. Also significant system-level throughput gain has been achieved using cross-layer adaptation policy compared with single-layer channel-dependent policy.

A study of energy efficiency of emerging rural-area networks based on flexible wireless communication is presented in [5]. Authors have given clear approaches to energy efficient PHY parameter adjustment and also added into consideration the notion of physically achievable modulation and coding schemes. In [11], a cross-layer adaptation scheme based on neural network is proposed that improves QoS by online adapting media access control (MAC) layer parameters depending on the application layer QoS requirements and physical layer channel conditions.

The problem of optimal rate control in wireless networks with Rayleigh fading channel is studied in [12], Dynamic programming based optimization technique is used to obtain the optimal rate control policy. Energy-efficient transmission techniques for Rayleigh fading networks are studied in [7], where the authors show how to map the wireless fading channel to the upper layer parameters for cross-layer design. An energy-efficient cross-layer design for a MIMO downlink SVD channel is given in [14]. In [4], authors has related the error vector magnitude (EVM), bit error rate (BER) and signal to noise ratio (SNR). They also present the fact that with such relationship it would be possible to predict or in cases substitute EVM in places of BER or even SNR.

Energy-efficient operation modes in wireless sensor networks are studied in [13] based on cross layer design techniques over Rayleigh fading channels using a discrete-time queuing model and a three-dimensional nonlinear integer programming technique. The authors in [10] have carried out joint optimization of the physical layer and data link layer parameters (e.g. modulation order, packet size, and retransmission limit). The problem of optimal trade-off between average power and average delay for a single user communicating over a memoryless block fading channel using information-theoretic concepts is investigated in [9].

1.2 Paper Organization

The paper is organized as follows. Section 2 describes system model including traffic and buffer models. Section 3 describes computation of transition matrix in WARP test-bed. In Section 4, we formulate the cross-layer design problem and discuss cost functions and constraints. Section 5 provides methodology of finding optimal policies and optimal costs for different objectives or QoS requirements using Lyapunov drift penalty algorithm. We discuss the results in Section 6 and conclude in Section 7.

2 System Model

We consider wireless transmission link, as shown in Figure 1, with a single transmitter and a single receiver system. This scenario can be found, for example, in cellular networks where we focus only on the transmission from a base station (or eNode B) to a single mobile receiver (or user equipment, UE).

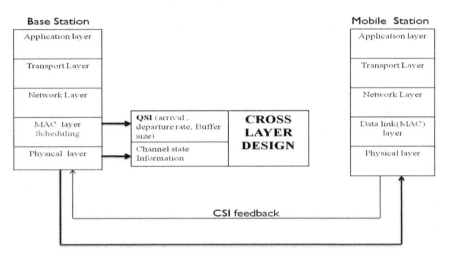

Figure 1 System model.

We assume that both the downlink and uplink transmissions are organized into radio frames. Time over which the source transmitter sends packets to the destination receiver is divided into a countably infinite number of discrete time-slots. A radio frame consists of several time-slots. The processing units are packets and blocks at the higher layer and at the physical layer, respectively. A downlink (or uplink) transmission block consists of multiple symbols and a packet is made up of multiple information bits. Packets from the higher layer application are stored in a finite size buffer at the transmitter. Based on the information of channel state, buffer state, BER and incoming traffic state, an adaptive modulation controller chooses the modulation scheme/constellation. The Adaptive Modulator unit takes corresponding number of packets from the buffer and modulates it with the chosen modulation scheme into symbols for transmission over Rayleigh diversity channel.

Let T_B denotes the duration of each time-slot. Therefore, the physical layer time-slot rate is $R_B = 1/T_B$ time-slots per second. We assume that the packets transmitted in one time-slot experience the same channel gain. Let b(n) denote the number of packets taken from the buffer for transmission and a(n) the number of packets arriving into the transmitter buffer from the upper layer application. Assume that the duration of a block is equal to N number of discrete time-slots and 'n' denotes the index of a particular. For example, a(n) denotes the number of packets arriving from higher layer at time-slot n.

The status of the system at any instant is described by one of the S possible system states. The job of a PHY-layer scheduler is to find the control action $\mu \in \{S_1, S_2, S_3, S_4, S_5, S_6,\}$ for all the time-slot. $S_n = 1, 2, ..., N$, where μ is the set of available actions and N (in number of slots) is the duration of communications. Our goal is to find an optimal stationary policy μ so that it translates the state into a corresponding optimal action, $\mu(s)$, We will discuss later how the action, which corresponds to transmission rate and/or power of the problem, can be selected when adaptation is made with only the physical layer, and also with cross-layer variables.

2.1 Buffer Modeling

Since in practical wireless transmitter, packets are stored in a buffer of finite-capacity, unlike most of the adaptive transmission methods in the literature, we assume that the size of the transmission buffer is finite and it can hold a maximum of B packets. Note that since the traffic is random in nature, the buffer may be empty sometimes and it may be full at some other times. If the buffer does not have enough space for all incoming packets, some packets will be dropped. Therefore, in this paper our goal is to bound both the queuing delay and the packet overflow. Particular value of these bounds depend on the QoS requirement of the application being considered. Let Q(n) denote the number of packets in the buffer at time slot n, therefore, the state space of the buffer's packet occupancy can be expressed as Q(n) = {0, 1, 2, ..., B}. Higher layer traffic produce arrival rate a(n), departing packets rate b(n) which is the function of channel, power.

2.2 Traffic Modeling

Usually, wireless network traffic is bursty, correlated and randomly varying. The Markov modulated Poisson process (MMPP) model, wherein, at any state, the incoming traffic is Poisson distributed A packets may arrive according to the Poisson distribution with average arrival rate λ_i packets per time-slot. In each time-slot, the transmitter selects b(n) packets for transmission over the wireless channel. In each time slot modulation schemes are chosen based on the cost function of the algorithm. Since the number of arrived packets a(n) and the number of packets chosen for transmission b(n) are randomly varying, the buffer occupancy fluctuates between 0 to B, where B is the storage capacity of the buffer. The buffer state at time-slot n can be given as

$$Q(n) = Q(n-1) + a(n) - b(n) \tag{1}$$

3 Warp Testbed

In this work, realistic Channel State information is obtained by using the WARP [16] test-bed. Total received signal strength is dependent both on distance and fading. We have assumed that the distance remains unaltered during the time period of interest, and hence we can just rely on fading to capture the variations in signal strength. However in situations where the above premise does not hold true one can combine this fading-based Markov chain model with mobility to model signal strength fluctuations.

The channel state partitioning can be done in different ways, but the equal probability method, where all the channel states have the same stationary probability, is the most popular in literature, because it offers a good trade-off between the simplicity and the accuracy for modeling a wireless fading channel. We denote the channel states by $C_k = \{C_1, C_2, C_3, C_4, C_5, C_6\}$, where the state is said to be in C_k when the gain lies between γ_{k-1} and γ_k as shown in Table 1.

3.1 Transition Matrix

The Markov chain transition matrix is obtained via an empirical approach, in which the Markov chain transition matrix is calculated by directly measuring the changes in signal strength. The transition matrix depicting SNR variation can be determined by collecting received signal strength measurements using the WARP test-bed and calculating the EVM values and then determining the transitions from one state to the other.

The transition probabilities $P_{ci,cj}$, c_i, c_j $\forall$ C_k were determined based on SNR variations which are obtained directly from EVM. The state of the channel can be estimated at the receiver and the information can be fed back to the tansmitter. When the perfect channel state information is available at

Table 1 Channel states

Channel State	SNR (dB)
C_1	3.0–5.0
C_2	5.0–8.0
C_3	8.0–10.5
C_4	10.5–14.0
C_5	14.0–18.0
C_6	Greater than 18.0

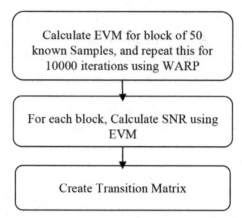

Figure 2   Flow chart for determining transition matrix.

the transmitter befre the transmission decision is taken, we usually refer to the channel as fully observable.

The transition matrix is determined in this work by performing signal strength measurements at the receiver for experiments conducted over fading channel. The first step in framing transition matrix is to calculate EVM values for each block with assumption that each block consist N samples. EVM can be calculated by using following expression [4],

$$EVM = \left[\frac{\frac{1}{T} \sum_{t=1}^{T} |I_t - I_{0,n}|^2 + |Q_t - Q_{0,n}|^2}{\frac{1}{N} \sum_{n=1}^{N} \left[I_{o,n}^2 + Q_{o,n}^2 \right]} \right]^{\frac{1}{2}} \quad (2)$$

$$SNR = \left[\frac{\frac{1}{T} \sum_{n=1}^{T} |I_t^2 + Q_t^2|^2}{\frac{1}{N} \sum_{n=1}^{T} \left[n_{I,t}^2 + n_{Q,t}^2 \right]} \right] \quad (3)$$

where

I_t, Q_t – Received symbol at t'th instant

$I_{0,t}, Q_{0,t}$ – Transmitted symbol at t'th instant

$I_{0,n}, Q_{0,n}$ – N Unique ideal Constellation Points

From above equations it is clear that, SNR is inversely proportional to square of EVM.

$$\text{SNR} \approx 1/\text{EVM}^2 \qquad (4)$$

Therefore, we have the sequence of SNR values from which we can obtain SNR transition matrix using hidden Markov model.

The subsequent step is to determine the number of transitions from each state to the others by observing the sequence of states. For example, suppose there are 6 states in all and that the sequence of states is $\{\ldots\ldots 2, 4, 6, 2, 4 \ldots\ldots\}$. The subsequence $\{2, 4\}$ means that we increment the number of transitions from state 2 to state 4 by one. The next transitions are from states 4 to 6, 6 to 2 followed by another transition from 2 to 4. Once all the transitions have been considered, we use the relative values of the number of transitions from state i to state j for all states j, to determine the empirical transition probabilities from state i to all states j, P_{ij}

A stochastic matrix (also termed transition matrix), is a matrix used to describe the transition of SNR between various states, in the present case SNR variation in a Rayleigh Channel. The transition matrix derived based on our experiment is as follows:

$$P_{ij} = \begin{array}{c} \\ S_1 \\ S_2 \\ S_3 \\ S_4 \\ S_5 \\ S_6 \end{array} \begin{bmatrix} S_1 & S_2 & S_3 & S_4 & S_5 & S_6 \\ 0.3651 & 0.2262 & 0.1710 & 0.1478 & 0.0817 & 0.0082 \\ 0.2286 & 0.2388 & 0.2105 & 0.1729 & 0.1289 & 0.0204 \\ 0.1390 & 0.2548 & 0.2221 & 0.1830 & 0.1617 & 0.0395 \\ 0.0592 & 0.2339 & 0.2294 & 0.2244 & 0.1893 & 0.0638 \\ 0.0092 & 0.1564 & 0.2669 & 0.2368 & 0.2215 & 0.1092 \\ 0.0001 & 0.0527 & 0.2109 & 0.3018 & 0.2491 & 0.1855 \end{bmatrix}$$

We observe from the above that the total variation is small, which implies that the distributions are close to each other. i.e., probability of SNR being in same state for next slot is higher than probability of changing to adjacent state in next time slot.

Based on the system state the corresponding cost functions are calculated and actions are taken accordingly. The possible actions are shown in Table 2.

These typical values have been used directly as possible action for traditional adaptive scheme. In our paper, we are taking Queue State information along with channel states to take optimum action.

Table 2   Possible actions based on channel states

Actions	MCS
S_1	BPSK (1/2)
S_2	QPSK (1/2)
S_3	QPSK (3/4)
S_4	16QAM (1/2)
S_5	16QAM (3/4)
S_6	64QAM (2/3)

4 Proposed Problem Formulation

4.1 Energy Considerations

In digital communications, as a general rule, energy consumption is lowered by either shortening transmission time or lowering transmission power. Higher bitrates lower the transmission time, but are sustainable only when the power is high enough to result in sufficient SNR. Thus, unless we allow for data to be dropped, a tradeoff between the time and the power exists. The theoretical relationship between bit rate and transmission power is given by Shannon's formula, which defines the boundary for the channel capacity. Since the formula does not provide a means to achieve the boundary bitrates, a theoretical solution can be practically infeasible. Moreover, in theory, the transmitter power is usually analyzed in isolation, while in reality the transmitter needs supporting hardware, which has non-zero power consumption.

4.2 Energy Efficiency

Considering an OFDM kind of multi-carrier transmission scheme in the physical layer, with some approximation (discarding the guard intervals), we can consider the subcarriers individually, and for each of them Shannon's formula defines the maximum achievable bitrate as:

$$R_i = W \log_2 \left(1 + \frac{\gamma_i}{\Gamma}\right) = W \log_2 \left(1 + \frac{p_{Tx,i}\, g_i}{N_o W \Gamma}\right) \tag{5}$$

where,

W represents the bandwidth occupied by a single subcarrier

γ_i represents signal-to-noise ratio

g_i channel gain

$P_{Tx,i}$ transmission power at the ith subcarrier

N_0 represents power spectral density of white Gaussian noise

Γ SNR gap

Total maximum data rate for k sub carriers is

$$R = \sum_{i=0}^{k} R_i \tag{6}$$

Total energy consumed by a bit is

$$E_{Txb} = \frac{P_{Tot}}{R} \tag{7}$$

where, P_{Tot} – Total Power

In data transmission significant part of the energy goes to the transceiver circuit power (P_{TC}), which takes into account the consumption of device electronics, such as mixers, filters and DACs, and is bitrate independent.

With a non-zero P_{TC} the energy consumption is:

$$P_{Tot} = \sum_{i=0}^{k} P_{Tx,i} + P_{TC} \tag{8}$$

The bitrate used in the calculations represents an upper bound. In physical systems the choice of MCS determines the actual bitrate. This bitrate is below the optimal for the given SNR, but is equal to the optimal for a channel with an SNR lower by a factor Γ. This factor is called the "SNR gap" and depends on the MCS used, as well as the desired bit error rate (BER). The energy per bit now becomes:

$$E_{Txb} = \frac{\sum\limits_{i=0}^{k} P_{tx,i} + P_{TC}}{\sum\limits_{i=0}^{k} W \log_2 \left(\frac{p_{Tx,i}\, g_i}{N_0 W \Gamma} \right)} \tag{9}$$

where, g_c = coding gain

$$\Gamma = \frac{log\left(\dfrac{P_e}{0.2}\right)}{g_c}$$

$$p_e = \frac{4}{log_2 M} Q\left\{\sqrt{\frac{a_\gamma \log_2 M}{(M-1)}}\right\}$$

and P_e is Probability of error.

4.3 Problem Description and Formulation

At each time-slot *n,* the scheduler chooses an action depending on the current system states. A decision rule denoted with μ specifies the action at time-slot. Decision rules for all time-slots over which communications takes place, constitute a policy of the problem. We consider a countably infinite horizon (i.e., horizon $H \to \infty$) problem, where our objective is to optimize long term average expected cost for different goals to be achieved. In this paper, we are to minimize the long term average transmission power under specified long-term average buffer delay and packet overflow. Let Π denote the set of all admissible policies π, i.e., the set of all sequences of functions $\mu = \{\mu 1, \mu 2, \cdots\}$ with $\mu_n: S_n \in S$ where S_n denotes the set of actions possible in state S. The cost function for a stationary policy is denoted by G_p. The objective of our cross-layer adaptation problem is to find the optimal stationary policy μ such that, the following objective is achieved.

$$\begin{aligned}
\text{Minimize} \quad & G_p = E_{tot} \quad &(10)\\
\text{subjected to} \quad & G_d \le G_{dth}\\
& Ge \le G_{eth}\\
& G_o = 0
\end{aligned}$$

where, G_{dth} and G_{eth} are the maximum allowable average delay and maximum allowable probability of error, respectively. $G_P(S_n)$ is the immediate transmission power cost at time slot n for action S_n. The long-term average expected queueing delay cost, G_d and packet overflow cost, G_O can be expressed in terms of the buffer backlog and the number of packets arriving per slot etc.

4.4 Cost and Constraints

4.4.1 Transmission power cost

Transmission power cost in a particular time-slot is the actual transmitter power used for transmitting packets. Suppose, our target is to keep average BER of the transmission the same irrespective of modulation scheme and channel state. The BER requirement can be specified from the application being handled. For a certain channel state and action S_i, and with a fixed specified average BER P_e, for all channel states, the power cost G_p is estimated with appropriate BER expression or using with instantaneous received SNR γ.

4.4.2 Queueing delay cost

Delay is an important parameter to consider for communication systems involving transmission buffers. The maximum tolerable packet delay for a particular system depends on the QoS requirements of the application being handled. For example, real-time traffic must have very low delay. For this traffic, the received packets are useful only when the strict delay requirements are maintained by the scheduler. On the other hand, best-effort traffic is not real-time and is quite tolerable to delay. The delay experienced by a packet is composed of buffer-queuing, encoding, propagation and decoding delay. In this paper, we consider only buffer delay, since encoding, propagation and decoding delay are usually fixed and are negligible compared to buffer delay. The average packet delay in the buffer is related to the average buffer occupancy via Little's theorem, as follows:

$$G_o(n) = \frac{Q(n)}{a(n)} \tag{11}$$

where, a(n) is the instantaneous packet arrival in slot n and Q(n) is number of packets (backlog) present in queue at time slot n.

4.4.3 Packet overflow cost

When the buffer is nearly full and the empty space is smaller than the number of packets arrival, packet overflow happens. Suppose, current buffer state is Q(n) and scheduler takes b(n) packets from the buffer. Therefore, buffer can accommodate r(n) = B − Q(n) + b(n) arriving packets in the current time-slot. Now, if arriving packets a(n) in particular traffic state is larger than r(n), (a(n)-r(n)) packets will be dropped with probability 1.

Therefore, packet overflow rate, for buffer state Q(n), traffic state f(n) and action S_n can be expressed by,

$$G_o(n) = \sum_{a(n)} \phi(a(n), (B - Q(n) + b(n)))^* P(a(n)) \qquad (12)$$

where (x, y) is a positive difference function, which returns the difference of x and y when $x > y$, and it returns 0 when $x \le y$. $P(a(n))$ is the probability of arrival rate a(n) at time slot n.

4.4.4 Delay constraint

Delay is an important parameter to consider for communications systems involving transmission buffers. The maximum tolerable packet delay for a particular system depends on the QoS requirements of the application being handled. We should maintain the delay at each slot to be less than the maximum permissible value (∼Buffer delay threshold).

$$G_D(n) \le G_{Dth}$$
$$\text{Let } g_d = G_D(n) - G_{Dth}$$
$$\text{then } g_d \le 0$$

and hence, we should maintain g_d as a negative value.

4.4.5 Error constraint

Our target is to keep average BER of the transmission the same irrespective of modulation scheme and channel state. The BER requirement can be specified from the application being handled. We should maintain probability of error less than some typical value based on QoS

$$G_e(n) \le G_{eth}$$
$$\text{Let } g_e = G_e(n) - G_{eth}$$
$$\text{then } g_e \le 0$$

we should maintain g_e as negative value.

4.4.6 Overflow constraint

We assume that the size of the transmission buffer is finite and it can hold a maximum of B packets. Note that since the traffic is random in nature, the buffer may be empty sometimes and it may be full other times. If the buffer does not have enough space for all incoming packets, some packets will be dropped. It may require retransmission which causes increase in energy. So our goal is to maintain zero overflow.

$$g_o = G_o(\text{n})$$
$$\text{i.e., } g_o = 0$$

5 Stochastic Optimization

5.1 Drift-Plus-Penalty Algorithm

The Drift-Plus-Penality algorithm is used to minimize the objective function on energy consumption subject to the constraints defined in our problem, that is, the Penalty functions whose time average should be minimized. Hence our problem as defined before is,

Min G$_D$ = E$_{tot}$

Subject to:

- $g_d(\text{n}) \leq 0$
- $g(\text{n})_e \leq 0$
- $g_o(\text{n}) = 0$

5.1.1 Virtual queue

For each constraint i in $\{1, ..., K\}$, virtual queue with dynamics over slots n in $\{0, 1, 2, ..., N\}$ is given as follows, [15]:

Delay:

$$Z_D[n + 1] = \max(Z_{D(n)} + g_d(n), 0) \tag{13}$$

BER:

$$Z_e[n + 1] = \max(Z_e(n) + g_e(n), 0) \tag{14}$$

Overflow:

$$H_o[n + 1] = H_o(n) + g_D(n) \tag{15}$$

where Z_D, Z_e, H_o are Lyapunov parameters used for creating virtual queue.

By stabilizing these virtual queues ensures the time averages of the constraint functions are less than or equal to zero, and hence the desired constraints are satisfied.

5.1.2 Lyapunov function

To stabilize the queues, the Lyapunov function L(n) is defined as a measure of the total queue backlog on slot n:

$$L(\theta(n)) = \frac{1}{2} \sum_{k=1}^{k} Q_k(n)^2 \qquad (16)$$

Squaring the queueing equation results in the following bound for each queue

$$L(\theta(n)) = \frac{1}{2}\{(Q(n)^2) + Z_D(n)^2 + Z_e(n)^2 + H(n)^2\} \qquad (17)$$

5.1.3 Lyapunov drift

The Lyapunov drift is given below and is used with the penalty functions in order to identify the control action to be taken.

$$\Delta(n) = L(n+1) - L(n) \qquad (18)$$

The drift-plus-penalty algorithm takes control actions in every slot n to minimize Cost function. Intuitively, taking an action that minimizes the drift alone would be beneficial in terms of queue stability but would not minimize penalty. Taking an action that minimizes the penalty alone would not necessarily stabilize the queues. Thus, taking an action to minimize the weighted sum incorporates both objectives of queue stability and penalty minimization as indicated below.

Lemma [15]:

$$\Delta[\theta(n)] + VE\left\{\frac{y_o(n)}{\theta(n)}\right\} \leq B + VE\left\{\frac{y_o(n)}{\theta(n)}\right\}$$

$$+ \sum_{k=1}^{K} Q_k E\left\{a_k(n) - \frac{b_k(n)}{\theta(n)}\right\} + \sum_{l=1}^{L} z_l(n)E\left\{\frac{y_l(n)}{\theta(n)}\right\}$$

$$+ \sum_{j=1}^{J} H_j(n)E\left\{\frac{e_j(n)}{\theta(n)}\right\} \qquad (19)$$

where

$$B \geq + \frac{1}{2} \sum_{k=1}^{K} E\left\{ a_k(n)^2 - \frac{b_k(n)^2}{\theta(n)} \right\}$$

$$+ \frac{1}{2} \sum_{l=1}^{L} E\left\{ \frac{y_1(n)^2}{\theta(n)} \right\} + \frac{1}{2} \sum_{j=1}^{J} E\left\{ \frac{e_j(n)^2}{\theta(n)} \right\}$$

$$- \sum_{k=1}^{K} E\left\{ b_k(n) \frac{a_k(n)}{\theta(n)} \right\} \tag{20}$$

5.1.4 Cost function

The expression for the cost function can be obtained by using the above lemma, as

$$Cost = V^* g_p(n) + (1 - V)^* \{Q(n)^* [a(n) - b(n)] + g_D(n)^* Z_D(n)$$
$$+ g_e(n)^* Z_e(n) + H(n)^* g_o(n)\} \tag{21}$$

where V = 0.5 states that we are giving equal importance to objective (drift function) as well as penalty function. In this work, we estimate the cost function for all six possible states and select the best out of them for each slot thereby approaching the optimized solution.

5.1.5 Simulation results

In this work, the above described optimization approach is used considering the traffic state, buffer state and the channel state that is measured for an indoor scenario using the WARP SDR module as explained in Sections 2 and 3. The performance of the adaptation policies with respect to departure rate in relay based wireless transmission downlink system with a transmitter and a receiver is shown. This indicates how the energy of transmission (Energy per bit) varies for each time slot based on overflow and delay (QSI) and SNR (CSI).

The performance was observed for 1000 time slots to gain an understanding of the dynamics and the inter-relationships. The SNR variation and Queue backlog as function of the time slot index are shown in Figure 3 and the variation in the probability of error and transmission energy at different time slots are shown in Figure 4. From these plots we can clearly observe that whenever SNR goes low, energy consumption goes high but vice versa is not true for the same entities. This is because energy consumption not only

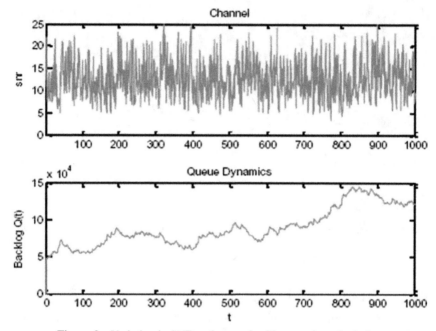

Figure 3   Variation in SNR and queue backlogs vs time slot index.

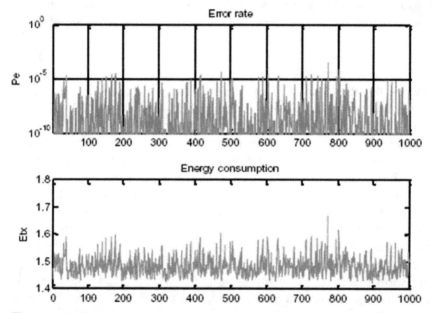

Figure 4   Variation in probability of error and transmitted energy vs time slot index.

depends on the SNR, it also depends on other constraints as defined in our problem.

It is further observed that around slot number 770, the transmitted energy is very high. This can be attributed to the increased queue backlog around that time and hence an increase in probability of error, which necessitates a corrective action.

The corresponding Lyapunov function and the Lyapunov drift at different time slot index are shown in Figure 5. We can notice that lyapunov drift is very high at initial slots. This is because of the sudden transition of Queue backlog from lower values (zero for initial slot) to higher values.

In Figure 6, we can observe that overflow is maintained zero throughout transmission due to the corrective actions being taken at each time slot based on our optimization. It is further observed from all these performance plots that the Cost function, though dependant on many constraints, is seen to be predominantly affected by Queue backlogs, which lead buffer delays.

The performance in terms of cost function and energy consumption are compared for the conventional CSI based adaptation approach and the crosslayer based approach proposed in this work and are shown in Figures 7 and 8, respectively. It can be observed that the proposed model improves the energy efficiency and also stabilize the cost function. Stability of cost function is achieved because the proposed model guerentees the stability of physical and virtual queues. The following estimates are made based on the above performances.

Total Energy Consumption (without cross layer approach) for 1000 slots = **1.4832 × 10³ mW**

Energy Consumption (with cross layer approach) for 1000 slots = **547.67 mW**

Total Cost (without cross layer approach) for 1000 slots = 4.01×10^{12}

Total Cost(with cross layer approach) for 1000 slots = 2.23×10^{12}

Thus the proposed approach is seen to significantly reduce the energy consumption by nearly 64% during the observation period, in comparison to the conventional adaptation strategy, in addition to stabilizing the cost function at a much reduced value.

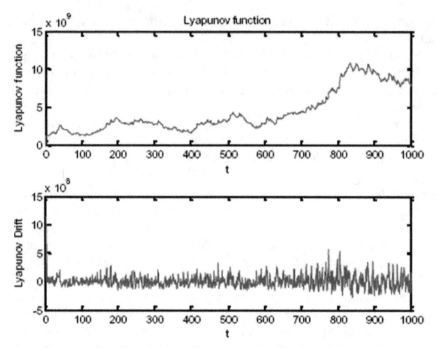

Figure 5   Lyapunov functions & lyapunov drift vs time slot index.

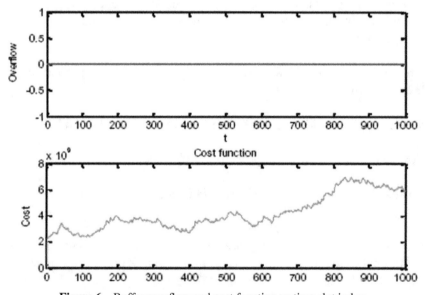

Figure 6   Buffer overflow and cost function vs time slot index.

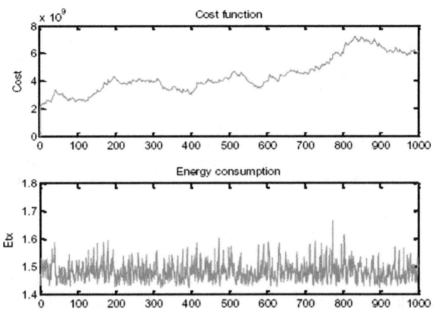

Figure 7   Cost function and energy consumption without cross layer.

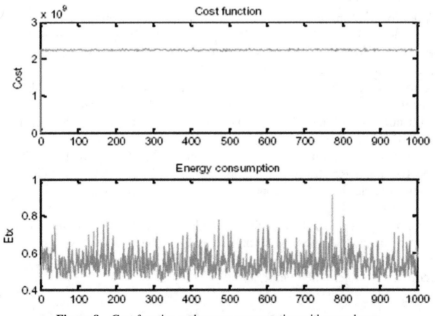

Figure 8   Cost function and energy consumption with cross layer.

6 Conclusion

In this paper, we have shown a possible strategy to design an intelligent packet transmission scheduler, which takes optimal transmission decisions using cross-layer information, based on Markov decision process formulations. We have discussed the method to compute the optimal policies when the channel states are perfectly observable and when they are partially observable. We have shown the benefits of a cross-layer policy over a single layer adaptation policy in terms of energy efficiency. By delaying packet transmissions in an optimal way, a huge amount of power can be saved for delay-tolerant data traffic applications. The amount of saving depends on factors such as the memory of a fading channel and the packet arrival rate. Such a cross-layer optimized packet transmission scheduling method will be a key component in future-generation green wireless networks.

References

[1] Ashok K. Karmokar, and Vijay K. Bhargava, "Performance of Cross-Layer Optimal Adaptive Transmission Techniques over Diversity Nakagami-m Fading Channels" IEEE Transactions Vol. 57, No. 12, December 2009.

[2] Antonio G. Marques, Carlos Figuera, Carlos Rey-Moreno and Javier Simo-Reigadas, "Asymptotically Optimal Cross-Layer Schemes for Relay Networks with Short-Term and Long-Term Constraints" IEEE transactions volume 12, no. 1, january 2013.

[3] Antonio G. Marques, Luis M. Lopez-Romos, Georgios B. Giannakis, Javier Ramos and Antonio J. Caamano "Optimal Cross-Layer Resource Allocation in Cellular Networks Using Channel and Queue-State Information" IEEE transactions volume 61, no. 6, july 2013.

[4] Rishad Ahmed Shafik Md. Shahriar Rahman AHM Razibul Islam On the Extended Relationships Among EVM, BER and SNR as Performance Metrics 4th International Conference on Electrical and Computer Engineering 2006.

[5] Veljko Pejovic and Elizabeth M. Belding. 2010. "Energy-efficient communication in next generation rural-area wireless networks." in proceedings of the 2010 acm workshop on cognitive radio networks (coronet '10). acm, new york, ny, usa, 19–24.

[6] Anand Seetharam, Jim Kurose, Dennis Goeckel, Gautam Bhanage "A Markov Chain Model for Coarse Timescale Channel Variation in an 802.16e Wireless Network" 2012 Proceedings IEEE INFOCOM.

[7] G. Li, P. Fan, and K.B. Letaief, "Rayleigh fading networks: a cross-layer way," IEEE Trans. Commun., vol. 57, no. 2, pp. 520–529, Feb. 2009.

[8] Ekram Hossain, Vijay k. Bhargava, Gerhard p. Fettweis "Green Radio Communication Networks" Cambridge University Press.

[9] R.A. Berry and R.G. Gallager, "Communication over fading channels with delay constraints," IEEE Trans. Inf. Theory, vol. 48, pp. 1135–1149, May 2002.

[10] H. Cheng and Y.-D. Yao, "Link optimization for energy-constrained wireless networks with packet retransmissions" Wiley Wireless Communications and Mobile Computing, vol. doi: 10.1002/wcm.996, 2013.

[11] C. Wang, P.C. Lin and T. Lin, "A cross-layer adaptation scheme for improving IEEE 802.11e QoS by learning" IEEE Trans. Neural Netw., vol. 17, pp. 1661–1665, Nov. 2006.

[12] J. Razavilar, K.J.R. Liu and S.I. Marcus, "Jointly optimized bit-rate/delay control policy for wireless packet networks with fading channels," IEEE Trans. Commun., vol. 50, pp. 484–494, Mar. 2002.

[13] X.-H. Lin, Y.-K. Kwok, and H. Wang, "Cross-layer design for energy efficient communication in wireless sensor networks" Wiley Wireless Communications and Mobile Computing, vol. 9, may 2012.

[14] D. J. Dechene and A. Shami, "Energy efficient quality of service traffic scheduler for MIMO downlink SVD channels," IEEE Trans. Wireless Commun., vol. 9, no. 12, pp. 3750–3761, Dec. 2010.

[15] Michael J Neely, Stochastic network Optimization with application to communication and queueing system, Morgon and Claypool Publications, 2010

[16] WARP Project, http://warpproject.org

Biographies

L. Senthil Kumar received B.E degree in Electronics and Communication Engineering from Coimbatore Institute of Engineering and Technology, Coimbatore, India, in 2009. He successfully completed M.E in Communication systems at College of Engineering, Anna University, Chennai, India in 2012. He is currently working toward the Ph.D degree at College of Engineering Guindy, Anna university, Chennai, India. His current research interests include Cross-Layer design, Green optimization in telecommunication and cooperative communication.

J. Vasantha Kumar is a M.E. Post Graduate student at the Department of Electronics and Communication Engineering, College of Engineering Guindy, Anna University, Chennai, India. He pursued his B.E in Electronics and Communication Engineering from Veltech multi tech Dr. RR Dr. SR Engineering College, Avadi, Chennai, Tamil Nadu, India. His field of interest is wireless communication and networks and Green Communication Networks.

M. Meenakshi Professor, Department of Electronics and Communication Engineering, Anna University Chennai, Guindy campus, Chennai - 600025; (e-mail: meena68@annauniv.edu), India, completed B.E (Honours), M.E and Ph.D in the years 1989, 1992 and 1998 respectively. She has been a faculty at Anna University Chennai, Since 1998. She is a member of IEEE, ISTE and Anna University Research gate. She has published nearly 35 national and international journal papers also more than 60 national and international conference papers in the field of Optical Communication & Networks, and Wireless Communications. Currently her research works are focussed on Green mobile communication, EMP effects, Wireless Body Area Networks, Radio over Fiber network Optimization, etc.

Setting Up a Low Cost Sustainable Telecommunication Infrastructure for Rural Communications in Developing Countries

Ram Krishna[1], Ravinder Ambardar[2] and S. Chandran[3]

[1]*Telecom Engineering Centre, Department of Telecommunications,*
Ministry of Communications & IT, Government of India, Janpath,
New Delhi-110001, India
[2]*Centre for development of Telematics, Department of Telecommunications,*
Ministry of Communications & IT Government of India, Chattarpur,
New Delhi-110030, India
[3]*Bharat Sanchar Nigam Limited (BSNL), Department of Telecommunications,*
Ministry of Communications & IT, Government of India, Janpath,
New Delhi-110023, India
Corresponding Authors: ddgfla.tec@gov.in; aravi@cdot.in

Received 2 Feb 2015; Accepted 5 March 2015;
Publication 29 May 2015

Abstract

This paper provides information on Government of India initiative to provide Broadband services to Rural India. In the absence of reliable grid supply in rural areas, deployment of alternate power system plays major role to power the Broadband systems deployed in rural areas. Service expansion requires fast site setup and low power consumption. This is being challenged by site acquisition and construction issues because of inconvenient access to many rural areas and energy supply that is often unstable or cannot keep up with demand. This contribution paper provides methodology based on the experience of the actual on-site techniques adopted in Broadband Network resulting in huge savings on operating expenses due to fuel consumption of diesel gen-sets and to optimize & harvest maximum renewable energy from available renewable sources in the rural areas having either no grid or poor grid.

Journal of Green Engineering, Vol. 5, 49–72.
doi: 10.13052/jge1904-4720.513

The power of broadband and ICTs to transform the lives of individuals economic & social wellbeing of nations is well recognized world over and so is the power of Optical Fibre as an underlying infrastructure for Broadband. As a result Broadband is critical and dear to hearts of nations. Transformation through Broadband however requires pre-requisites of various factors of its eco-system to be in place for its take off. The complexity of eco system in Indian scenario amongst other factors is characterized by vast linguistic, cultural, terrain, diversity, affordability and digital literacy. This would mean multipronged parallel efforts on many fronts on parts of government, industry and all stake holders to contribute towards development of ecosystem before the true benefits of Broadband can reach masses there by changing fortune of nation. Already some work has been done by various stakeholders in various areas which have a good bearing on off take of Broadband. The experience needs to be leveraged.

The important property that can make Information and Communication Technology (ICT) interesting for human is human welfare and development. ICT bestows upon humanity the ability to defy distance and time. Human development is quantified in the annual World Human Development Report of the United Nations as progress in health and education. A healthy nation means more productive labour and an educated nation means more creative labour. ICT has many facets. The most visible part is the bandwidth used for communication. Modern technology delivers gigabits through a fibre optic medium and several megabits through the wireless medium. A combination of the two technologies along with specialised devices often called routers and switches (equivalent to post offices and beat constables) can enable flow of gigabits of information from one village to another. The villagers can have access to high quality medical help, quality education, and relevant information pertaining to crops, fertilizers, entertainment, and access to the Internet as is enjoyed by their urban counterparts.

Around 70% of telecom towers are in rural areas, where grid connected electricity is not available and as a result, a very large chunk of the towers are powered by diesel generators which produce a total of 5.3 mnliters of CO_2 every year. Due to this high dependence on diesel, the operational costs of these cell sites increase drastically to about 200% more than those where grid power availability is regular. So operators are left with no other option than to look for alternate power supply solutions like wind power, solar power, hybrid, or bio-diesel solutions. With decreasing ARPU and increased opex, operators need a future-oriented wireless network solution to handle the challenges and boost profits. Instead of looking for green solutions for their energy requirements for the 0.35 mn plus towers, they looked for power efficient processes and more energy saving equipments for their entire network, and not just at tower sites. The major challenge today is also the high capex for the hybrid solutions, even though the ROIs are better. The telcos are looking out for the financial model, which is based upon opex.

1 Background

The purpose of this contribution is to develop recommendations that identify the general necessities on the requirements of rural area telecommunication infrastructure in developing countries. These recommendations may be helpful for equipment manufacturers to develop products or architect solutions, which are more appropriate for rural areas which are sparsely populated and under-served areas; and would help network operators/service providers in procurement process with business case to set up a low cost sustainable telecommunication infrastructure for rural communications in developing countries.

The power of broadband and ICTs to transform the lives of individuals economic & social wellbeing of nations is well recognized world over and so is the power of optical fibre as an underlying infrastructure for Broadband. The complexity of ecosystem in Indian scenario amongst other factors is characterised by vast linguistic, cultural, terrain, diversity, affordability and digital literacy.

2 Discussion

Fibre technology has something very interesting for economists. The concept of marginal cost deals with increased production with incremental investment, once the basic system is in place. In fibre technology, a mere two per cent incremental investment creates more than 100 per cent production capacity. This is mainly because when the fibre is laid, it has 6/12/24/48 cores inside and only one pair is put to use. The rest can be utilized as and when necessary to increase bandwidth or carrying capacity. That is the incremental cost. India has fantastic facilities to produce fibre optic cables in bundles up to 96 cores and beyond. *Fibre optic cable holds the key to a rural revolution. It is creating a multi-purpose infrastructure for the villages of India.*

India with its vast population of 1.21 billion (0.833 billion i.e. 69% rural population) and area of 3,287,240 sq. kms. with 29 diverse states and 6 union territories is administratively divided into districts (640), blocks

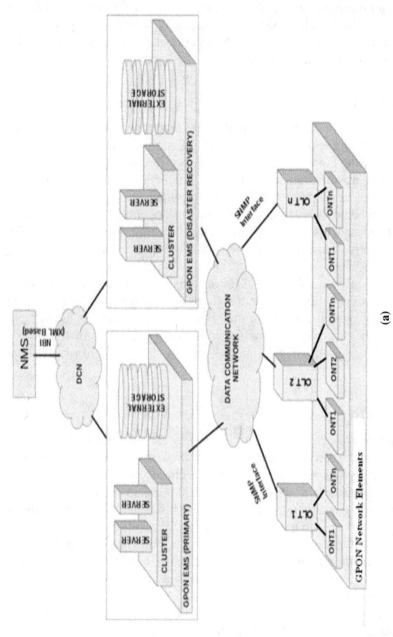

Figure 1   (a) GPON in the network to provide triple play services to the end customers.

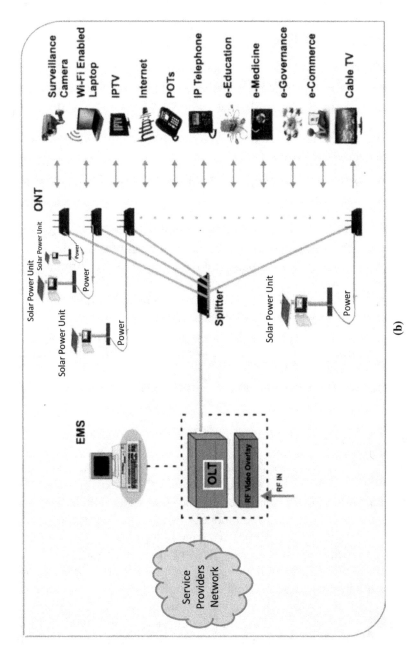

Figure 1 (b) The interconnection of various elements of GPON in the network.

(6382) and villages (6,38,619). There are 250,000 Gram Panchayats (Village councils) as part of administrative set up to govern these 6,38,619 villages i.e. on an average 2–3 villages per Gram Panchayat (hitherto referred as GP in the article). Even in urban India, Broadband is to realize its full potential, rural India is still far away. Factors which create urban-rural divide are (i) Difficult terrain & Scattered Population (ii) Lack of Infrastructure (Roads, Power etc.) (iii) Low income (iv) Higher capex and opexin rural areas.

The Communication technologies deployable in the rural areas need to meet the following requirements:

- Provide broadband connectivity at 100 Mbps or higher for community services.
- Be reliable and rugged.
- Require low power and be operational on alternative sources of energy such as renewable energy sources e.g. solar, wind and fuel cells.
- Have an attractive price/performance ratio for low cost transmission lines with sufficient bandwidth.
- Conform to international standards for service interoperability.
- Easy support for wireless and wired networks.
- Require easily available, low cost and end user friendly equipment.
- Preferably be maintenance free or utmost require low skill maintenance.
- Be easily and locally repairable.
- Do not demand mandatory operating environments like air-conditioning.

3 GPON as an Access Technology

Gigabit Passive Optical Network (GPON) technology can offer an excellent mix of triple play services to end users. It provides the backhaul as well as Access for provisioning of broadband services in rural areas. The indigenous GPON equipment developed by government body namely Centre for Development of Telematics (C-DOT) offers advantage in terms of appropriateness for Indian Environment, innovation, easy & fast deployment and local manufacturing. Current technology delivers a downstream data rate of 2.5 Gbps and an upstream data rate of 1.25 Gbps on a single fibre over a distance of 20 Km. One optical fibre from OLT can be shared up to 128 users. Signals from OLT to the ONTs are encrypted and then broadcasted to work station devices. Signals from the work station devices are then multiplexed back to the OLT.

4 National Optical Fibre Network (NOFN)

The NOFN project will build a strong middle-mile, but to build sustainable economic models around relevant e-services for the rural masses, there is a need to deliberate on the core and the last mile. The salient features of NOFN are:

- NOFN is being built using dark fibres leased from government owned three Public service units (or any other desirous Telecom operator) and laying incremental fibre.
- To offer interconnection at Block level and Gram Panchayat (GP) level.
- To offer guaranteed Bandwidth of 100 Mbps at Gram Panchayat.
- Uses technologies that are scalable, maintainable, observable & controllable meeting ground realities of diverse rural environment.
- NOFN to be operated and controlled by NMS centrally.

5 Use of Alternate Source of Powering

This technology has one of the lowest power requirements per customer. The customer premise equipments (ONTs in case of GPON) are low power devices which use between 8 to 15 watts of power. Since the electric power availability in rural areas is dismal, alternate use of powering the devices is required. These CPEs can be powered with solar energy by having solar panels installed near the end devices. In fact, the Wi-Fi terminals installed

Service layer:
(High speed internet, Video/voice Calling, Video Entertainment and e-services (Learning, Health, Retail, Banking, Governance)
State (28)/District(640) level
Core layer: Fibre Transport (Core Fibre to be provided by PGCIL, BSNL and RAILTEL)

PGICL DWDM	BSNL DWDM	RAILTEL DWDM

Block (6283) level
Middle mile layer: Fibre Access GPON
Gram Panchayat (2,50,000) level
Village(630619) Last mile layer: Service providers
Wi-Fi
To be provided by private telecom operators through non discriminatory access
End user devices (PC, Mobile phones, Tablet, Laptop, STB, Voip terminals, Retail PoS)

Figure 2 (a)   NOFN infrastructure.

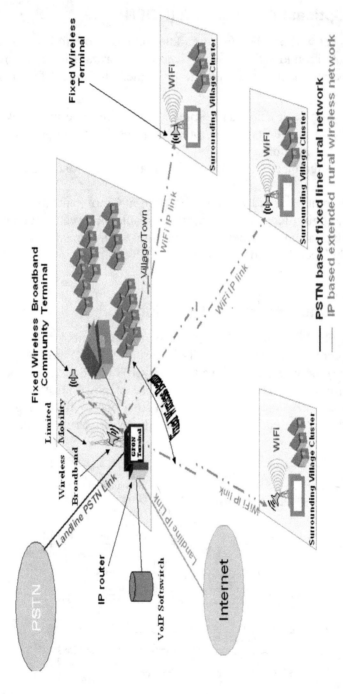

Figure 2 (b) NOFN rural applications.

for last mile connectivity can also be powered through solar solution. This is essential to keep the network up and working with uninterrupted services to the customers.

6 Lightning and Surge Protection for Photovoltaic (PV) Systems

PV systems being located and installed in exposed positions are subject to all conditions of weather. The measures to protect the sensitive electronic system components from failure due to lightning flashes and surges are absolutely necessary. Surges in PV systems may be caused by inductively or capacitively coupled voltages due to lightning discharges and switching operations in the upstream alternating current system. *The surge protective device combines a type1 lightning current arrester with a type2 surge arrester. It incorporates the proven Y circuit with the short circuit interruption (SCI) technology.*

On the DC side, a PV surge protective device is to be installed in each generator junction box/string combiner box. Surge protection devices shall have three-step d.c. switching element. This consists of a combined disconnecting and short-circuiting device with Thermo Dynamic Control and an additional melting fuse. The integrated parallel fuse disconnects the arresters safely from generator voltage in case of overload and allows for a safe and dead (arcless) replacement of the respective protection modules. The formation of a d.c. switching arc is prevented. The synergy of technologies applied in the Surge protection reduces the risk of protective devices being damaged due to installation or isolation faults in the PV circuit, clearly reduces the risk of fire at an overloaded arrester and puts it into a safe electrical state without interfering the operation of the PV system.

7 Core Network Considerations

NOFN would enable access and usage of several e-services such as healthcare, education, financial services, agriculture, e-governance, entertainment, etc,. The planned provisioning of 100 Mbps bandwidth through NOFN at the Gram Panchayat level translates into demand of 60 Gbps (average demand assuming 10 percent concurrency) per State Head Quarter. Further, as per various estimates, data growth is expected to increase around five times over a five year period. Current core capacities of service providers may not be adequate to cater to this demand. A parallel upgrade of core capacities may therefore, requires to be considered along with NOFN rollout.

8 Last Mile Considerations

While NOFN is a commendable step in bridging the urban-rural divide in broadband penetration, last mile access would be critical for realizing the policy objectives of inclusion and universal access. As we explore viable economic models for rural e-services, the critical need for the last mile becomes apparent. Middle-mile fiber layout enables affordable delivery of critical services at the Panchayat level through community-based service provisioning. However, efficient and viable delivery of certain community services would require extending connectivity to the last mile (for example, school premises in the case of education). Extending the last mile access to individual households would prove conducive to awareness and uptake, as rural users get accustomed to electronic delivery of essential services. This would require collaboration between the Government and private sector enterprises to work out strategies that make the proposition viable for all stakeholders. The last mile connectivity from Panchayats to adjacent schools, hospitals etc can be done either extending the fibre to those places or alternatively through wireless, if fibre is not a viable solution. Wireless solution based on Wi-Fi technology is the best option to provide services beyond panchayats to connect schools, hospitals, IT Centresetc in rural areas. Broadband wireless solutions with Wi-Fi technologies are available which work on license-free band. This technology can be used to connect remote areas where possibility of laying cable is difficult.

9 Economic Models for Sustainable Development

The model/approach followed is whole-seller/retailer model. Bharat Broadband Nigam Limited, a Special Purpose Vehicle (SPV) is a whole seller of bandwidth at 2.50 lakh Gram Panchayat level using which TSPs/ISPs can launch access services in conjunction with content/application providers. There will be two types of usage – Govt. to Customer (G2C) and Business to Customer (B2C). First use of NOFN to trigger broadband system through government lead uses e-governance, education in school, tele-medicines in hospitals, etc., to address the inadequate ratio of teachers to students, doctors to patients, police to citizens.

Providing relevant broadband-enabled services through a Public-Private-Panchayat ecosystem requires strong economic models with clear returns for each stake holder. While the state bodies would demand tangible social returns in the form of employment generation and skill-building for the rural population, private sector enterprises would seek long-term commercial viability.

There are certain basic prerequisites for creating practical economic models for rural areas. For instance, the availability of basic infrastructure, such as continuous power supply and uninterrupted connectivity, is imperative for delivery of e-services to rural consumers. Furthermore, building for these services is vital to ensure commercial viability of these initiatives. The availability of G2C services such as issuance of birth and death certificates, land records, Right to Information (RTI) services, etc. would attract people to the CSCs, thereby, giving them an opportunity to look at other low cost e-services such as e-learning modules, banking, etc. Economic models that involve delivering broadband to the household level also necessitate last mile connectivity – a subject which still awaits clear decisions and policies.

9.1 Economic Model I: Education

The private sector has also taken several initiatives towards improving the state of Indian education through ICT. Some cases in point are Microsoft India's Shiksha program and Sahaj e-Village's e-learning offering, described below.

9.1.1 ICT centers in Government schools set up by private enterprises

Delivery of quality school education in rural areas through ICT intervention is the main objective behind this model. It evaluates setting up ICT Centre(s) or digital classrooms in rural Government schools to deploy learning method-ologies such as audio-visual modules, animation and remote learning. The real value will be derived from virtual classrooms, using which, a teacher located in an urban area (district/block level) can conduct a class with school students located in rural schools. The job of supply, installation, maintenance of IT infrastructure and supply of learning content can be awarded to private education service providers under a Build Own Operate Transfer (BOOT) model (Figure 3).

9.1.2 Revenue potential

The Government of India will provide funds to the government primary schools to cover per student usage fee in lieu of connectivity through NOFN. The government primary schools at Panchayat level will pay per student usage fee to the private education service provider in lieu of digital learning set up to

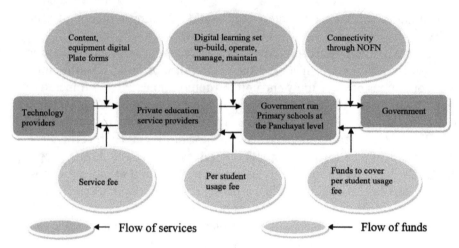

Figure 3   Economic model for digital classrooms in government schools.

educate the students. The private education service providers will share their revenue with technology providers for the provisioning of contents.

9.1.3 Vocational training delivery via CSCs

Skill development and employment generation through certified vocational training courses delivered remotely with the help of ICT is the main objective under the scheme. Job creation and the creation of a skilled workforce are of utmost priority for rural India. However, the lack of suitable teaching personnel and other infrastructure forms a key bottleneck. Using the NOFN infrastructure, it is possible to train rural youth through e-learning content spanning important skills such as basic accounting, BPO training, basic paramedical training, computer fundamentals, web designing, spoken English courses, computer hardware engineering, etc. Advanced courses such as MS-Office, CorelDraw, Tally, etc. can also be made available. Along with training, it is also important to be able to provide placement assistance to the trainees. Unlike primary education, the willingness to pay for such courses is relatively high, and this demand-pull can be used to create sustainable Economic models in vocational training. One of the possible delivery models for vocational e-training could be private training service providers partnering with the CSCs to deliver such content. The infrastructure in connected primary schools can also be leveraged for this service after school hours, with some additional manpower requirement. There is no economic support required from the government in this model, but the government can lend support by

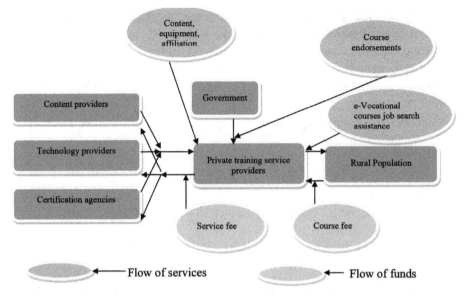

Figure 4 Economic model for vocational training and employment generation.

endorsing or certifying some of the courses offered through this model. Such endorsement is likely to enhance the credibility of these courses in the job market, and therefore, create demand (Figure 4).

Revenue Potential: The revenue earned by private training service provider by way of imparting e-vocational courses and job search assistance to rural population will be shared with content providers, technology providers, and certification agencies for contents equipment and affiliation. Government of India will endorse the courses to be run by private training service providers.

9.2 Economic Model II: Healthcare

Improving healthcare in rural India is one of the priority areas for the Government of India. The 12th Five-Year Plan intends to raise healthcare expenditure (Center and State allocation combined) from 1.3 per cent of GDP in Five Year 2012 to 2.5 per cent of GDP by Five Year 2017–27. The Planning Commission also aims to achieve Universal Health Coverage (UHC) by 2022–28 and the Government intends to achieve this in collaboration with the private sector (including NGOs and non-profits).

9.2.1 Commercial telemedicine centers owned and operated by private healthcare providers

Delivery of quality medical advice to rural patients at their doorsteps through Telemedicine in rural India is not a new concept. Many of the broadband pilots and trials under progress in the country are already using this alternative form of healthcare delivery. This model looks at widening access to good quality primary healthcare at the Panchayat level through telemedicine centers that can facilitate real-time two-way video calls between rural patients and doctors operating from urban hospitals.

The model should be able to remotely draw on the expertise of trained and qualified medical personnel available in urban hospitals.

- **Access issues:** The model should be such that rural patients get access to healthcare facilities nearer home.
- **Low ability to pay:** The model should not impose high additional usage charges on the patients.

All the above can be achieved through telemedicine – the exchange of medical information from one site to another via ICT tools using two-way video over a high-speed communication network.

Telemedicine centers are equipped with technology that enables access to specialist consultants situated in urban centers through real-time video conferencing techniques. This addresses the problem of low doctor to patient ratio in rural areas and also makes it possible for rural patients to access specialist advice. The staff manning the unit are also capable of providing basic medical treatment and procedures, eliminating the need to travel far to access basic care. The telemedicine centers can either be mobile (e.g. mobile vans) or stationary (housed in a brick and mortar premise, e.g. the Primary Healthcare Centers, or CSCs, Figure 5).

Revenue Potential: The revenue earned by primary health care service provider in the form of consultation fee from the patient for telemedicine services will be shared by technology providers as service fee towards equipment and technology platform provided by technology providers and other alliance partners.

9.2.2 Rural entrepreneur provides telemedicine services in collaboration with private players

A slight variation to the model described above would be a rural entrepreneur acting as the primary interface of service delivery. This model assumes

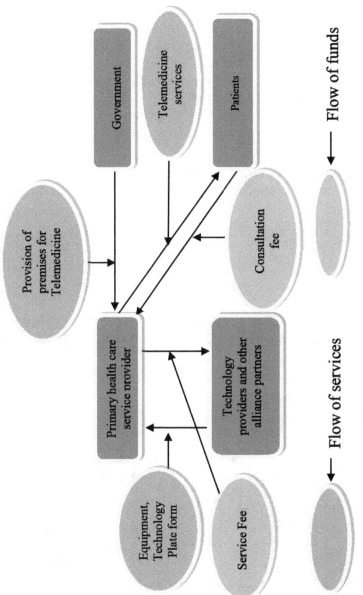

Figure 5 Telemedicine service through a private health care service provider.

leveraging the existing CSC infrastructure. The Village Level Entrepreneur (VLE) running a CSC can tie up with a private healthcare service provider to offer telemedicine services. The incremental cost of setting up a telemedicine unit in a CSC will be lesser compared to the complete build-out cost in model 1 because many of the cost elements are already in place at the CSC – e.g. the space, the basic computer and peripherals, manpower, power back up, web-cams, furniture, etc. The acceptability of the service is likely to be higher, as it is provided by a rural entrepreneur. Leveraging existing infrastructure to offer more acceptable telemedicine services, while building skills, encouraging rural entrepreneurship, and generating employment are the main propositions (Figure 6).

Revenue Potential: The revenue earned by primary health care service provider in the form of consultation fee from the patient for telemedicine

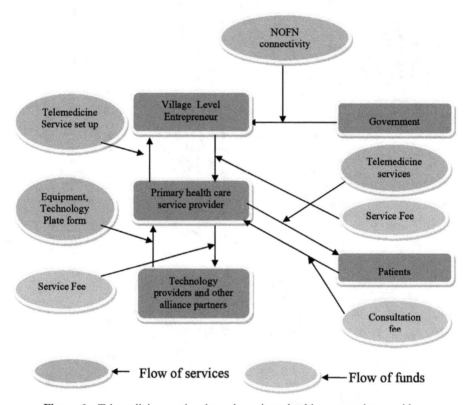

Figure 6   Telemedicine service through a private health care service provider.

services and service fee earned from the village entrepreneur towards the telemedicine service setup will be shared with the technology providers as service fee towards equipment and technology platform provided by them.

9.3 Economic Model III: Banking

Nearly 70 percent of the Indian population resides in rural areas; however, approximately 61 percent of the rural population remains unbanked. The banking network in the country has not yet been able to penetrate into many parts of rural India. While close to 31 percent of villages (about 200,000 villages) have at least one branch of a commercial bank, about 46 percent of the rural households still do not avail of banking services.

9.3.1 Banks tie up with ICT-enabled post offices
Leveraging Post Office infrastructure to offer banking services in rural areas is the objective for providing banking services. This model harnesses the strong presence of Post Offices in rural India to drive ICT-enabled financial inclusion. As of December 2012, the 169 banks in India – including 82 regional rural banks – had a branch network of 100,277, of which only about 37 percent were in rural areas. In contrast, India Post has 154,822 post offices, of which 139,086 (or ~90 percent) are in the rural areas. Using this significant coverage to offer banking services could boost the Government's financial inclusion initiatives. Similar models have been implemented internationally and could be explored in India, given that certain prerequisites to the model, such as banking license for India Post and connectivity at post offices, are already under consideration. In this model, bank appointed Banking Agents (BAs) in the villages access the nearby post offices (as against bank branches) to carry out cash transactions (e.g., deposit the money collected from account holders). The post office would, in turn, carry out settlement with banks (for which they would require a banking license, Figure 7).

Revenue Potential: The revenue earned by the bank from government in the form of service fee for delivery of welfare payments will be shared with the postal department in the form of share of the transaction fee based on the value of the transaction in lieu of Infrastructure for banking services to BC's and with technology providers in the form of service fee towards software and card readers to handle basic banking services. Banking agent will be getting account opening fee from the account holders for banking services and commission/salary from the banks for acting as bank representative in remote areas for extending the banking services to the account holders.

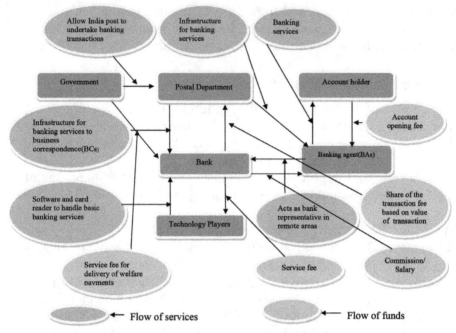

Figure 7   Bank led Economic model using the post office network.

9.3.2 Third party-led BC model

This model is led by a third-party service provider which acts as the Business Correspondent for the bank(s). This model is already under operation in India and could be extended to unconnected regions with the advent of NOFN. The third-party BC recruits and trains Banking Agents (BAs) to offer banking services in remote areas through handheld Point-of-Transaction (POT) terminals provided by the third-party service provider (Figure 8).

Revenue Potential: The revenue earned by the bank from government in the form of service Fee for delivery of welfare payments will be shared with the third party service providers/business correspondence as onetime fee for customer acquisition and for transaction towards acting as a bank representative in remote areas. The third party service providers/business correspondence will make payment as service fee to technology players towards software and card reader to handle basic banking transactions and payment to banking agent in the form of commission/salary towards customer reach. Banking agents will also get account opening fee from the account holders for banking services.

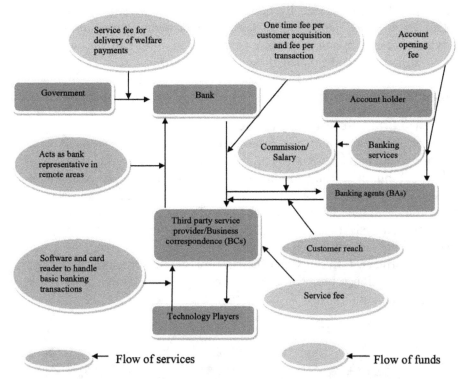

Figure 8 Third party led economic model harnessing a BC network.

9.4 Economic Model IV: Agriculture

Agriculture is the principal source of livelihood for 58 percent of the Indian population and hence, plays a major role in the overall development of the country. ICT plays a vital role in uplifting the agricultural community by providing technology enablers that enhance farm productivity and lead to increased income levels of farmers. Initiatives such as e-Choupal, e-Kutir, and e-Krishi have demonstrated that deployment of ICT tools can increase farmers' income by up to 400 percent. The infusion of ICT in agriculture not only benefits farmers but also creates an entire commerce ecosystem for agro input providers, agro product purchasers and consumer goods companies. More importantly, ICT reduces information asymmetry (e.g. price asymmetry, lack of awareness with respect to latest technological trends, seed types, and demand for various crops) which is crucial to equitably distributing the fruits of development among all sections of the agriculture community. Acknowledging the fact that timely access to agricultural information is one of the major challenges faced by Indian farmers, the Government of India launched 'Kisan Call Centers' with an objective of providing agricultural information to farmers. While this is a commendable step, information asymmetry could be further reduced by providing additional services such as real-time alerts, e-commerce etc. This is a concept which has been experimented successfully in various pilots undertaken by the private sector in India.

9.4.1 Rural entrepreneur-led model

This model involves training and empowerment of rural Entrepreneurs who work closely with farmers towards real time access to agricultural information through ICT, building rural entrepreneurial skills in the process. Social businesses such as eKutir have proved that a scalable model can be built on these lines and the various relevant ICT offering – applications for seed selection soil analysis, crop planning tools, etc., – can be offered. The social businesses, with their affiliation networks, can source the necessary tools and content for farmers. Offerings can include low-cost information on seeds, fertilizers, cultivation, types of crops grown, crop prices, and other key aspects of agriculture (Figure 9).

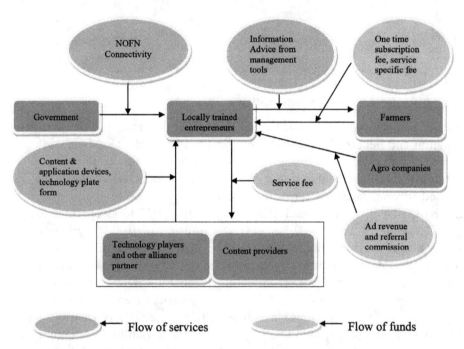

Figure 9   Economic model for rural entrepreneur-led agri. knowledge hub.

Revenue Potential: The revenue earned by the locally trained entrepreneurs from the farmers for providing information advice from management tools and earned from the agro company as advertisement revenue and referral commission will be shared with technology and content providers towards provisioning of contents & application, devices and technology plate form.

10 Proposal

Keeping in mind about the ensuing boom of Broadband penetration expected in rural areas, it is hereby submitted that development of recommendation for defining general aspects/parameters for Telecom infrastructure in rural areas of developing countries can be considered. This may require study of power consumption scenarios in the equipments being deployed in rural areas of various developing countries. Some of the requirements which could be incorporated may be:

1. Providing details of energy efficient Broadband equipment which may help to bring down cost of equipment being deployed in rural areas.
2. Providing typical advantages of moving towards use of Hybrid Renewable energy Sources Including wind energy, solar energy and fuel cells for powering Broadband equipment in the rural areas.
3. Providing typical gains by using wireless backhaul systems for access transport specially in rural environment.

References

[1] The Journal of the Optical Society of America.
[2] European Conference on Optical Communications.
[3] FTTH Council Asia Pacific.
[4] Creating viable business models for inclusive growth through the National Optical Fiber Network-Pradeep Udhas, KPMG in India and Confederation of Indian Industry (CII).
[5] Lightning Protection Guide-DEHN International.

Biographies

R. Krishna, an M.Tech. (Electronics) is serving in Department of Telecommunications, Ministry of Communications, Government of India as Deputy Director General (Fixed Access) and is responsible for formulation of standards on Telecom Equipment to be deployed in Indian Telecom Network on fixed line access technologies which includes the broadband access Technologies on copper and fibre. He has also been responsible for validation of GPON access equipment for Technology approval developed by Center for Development Telematics (C-DOT) in India which is being deployed In Indian Telecom Network for provisioning of Broadband services in Rural Areas. He is also working on standardization of Telecom Equipment to be performance and energy assessed and certified Green passport. He has varied experience of working in various disciplines for installation, Testing and certification of Telecom Equipment on Digital Coaxial, Satellite, GSM and Optical Fibre technologies deployed in Indian Telecom network.

R. Ambardar has done his Engineering in Electronics from MNNIT, Allahabad and is presently working in C-DOT- a premier telecom technology centre of Government of India, as Group Leader. He is currently involved in various activities like product specifications design, validation and testing, field support and customer interactions in the area of optical communication. He is responsible for successfully deploying India's first indigenously developed GPON system in India at Ajmer, Rajasthan in 2010 and later on successful field trial completion of GPON system in the NOFN netwok. He has been a member of core group for NOFN whose task was to design and deliberate on the Architecture of the network for providing broadband connectivity to 2,50,000 village Panchayats in the country. He was also the Member of Technical Committee formed by DeitY for devising strategic plan for providing G2C services on NOFN and member of NII 2.0 architecture committee setup by DeitY for easier implementation and maintenance of e-governance for state and central line departments. He is also the member of Technical Audit methodology committee set up by DOT to discuss on the recommendations of preferential market access (PMA) for electronic products in the telecom sector.

S. Chandran had served as Assistant General Manager in Bharat Sanchar Nigam Limited (BSNL), a premier telecom service provisioning unit of Department of Telecommunications, Ministry of Communications, Government of India. He has been associated with the validation and acceptance testing of various kind of telecom equipment being inducted in BSNL's Telecom Network for service provisioning. He has also been responsible for formulation of validation schedules, test schedules and test procedures for testing and certification of Core, Infra and other sub-elements of different telecom technologies both TDM and IP based for induction in the BSNL's telecom network prior to commercial launch. He has also been a key person to extend the technical support in formulation of standards on Telecom Equipment to be deployed in Indian Telecom Network on fixed line access technologies by TEC which includes the broadband access Technologies on copper and fibre. He has varied experience of working in acceptance, testing and certification of Telecom Equipment on GSM and Optical Fibre technologies deployed in BSNL's Indian Telecom. He has been conferred with the many awards for his meritorious services in acceptance, testing and certification of telecom equipment for use in BSNL telecom network.

Scheduling BTS Power Levels for Green Mobile Computing

Hemalatha M.[1], Prithiviraj V.[2] and Jayasri T.[3]

[1]Department of ECE, Narayana Engineering College, Nellore,
Andhra Pradesh, India
[2]Rajalakshmi Institute of Technology, Chennai, Tamil Nadu, India
[3]Tata Consultancy Services, Chennai, Tamil Nadu, India
Corresponding Authors: mkhema@yahoo.com; {profvpraj;
jayasri5591}@gmail.com

Received 2 Feb 2015; Accepted 5 March 2015;
Publication 29 May 2015

Abstract

The world has become sophisticated and man started depending upon various things in a large part of everyday life to enhance daily way of living which includes the most useful gadgets. These resources have made life easier, but there is a lot of concern over the possible health risks due to these radiations. Considering the objectives of Green Mobile Computing, this work proposes to dynamically allocate BTS's transmission power level according to the requirement of number of users in a particular radio cell. The number of users is more during the daytime when compared to night time, so the power level can be dynamically reduced to overcome the hazardous effects of radiation. A scheduling algorithm for switching the transmission power level at BTS, based on the output derived from neural network, which is trained with historical data collected from local authorities to learn the population pattern and assign corresponding power levels. The scheduling algorithm with Artificial Neural Network (ANN) gives reduced power consumption, low interference and shortened radiation exposure.

Journal of Green Engineering, Vol. 5, 73–84.
doi: 10.13052/jge1904-4720.514

Keywords: Green Mobile Computing; EM pollution; BTS; Scheduling; PoP.

1 Introduction

Wireless technology plays major role in the field of telecommunication where data and voice are frequently exchanged through energy content offered by electromagnetic spectrum. Sophisticated hand on services offered by telecommunication system has increased the number of users. Thus the base stations deployed for managing these load acts as a major artificial source for generating EM radiations. For the major part of the world, to go even a day without coming into contact with any of these devices is impossible. However, most people do not realize that the adverse effects of deterioration of man's living are caused by the electromagnetic radiations that are emitted by all these electronic devices. Though the awareness of this fact is ignored by many, by having a detailed study of this type of radiation and its negative impact and its consequences for human living. The overwhelming research that has been done on this matter provides a substantive evidence and undeniable facts that electromagnetic radiations that has been emitted from high tech devices can be very harmful. Harmful radiations has some other added disadvantages as the link between these radiations and various forms of illness has been revealed through many studies.

As the number of mobile users increased tremendously the energy level of radiations encompassing the environment has also increased and resulted in health hazards. So in order to overcome this, the transmitting power level at BTS is adjusted efficiently based on the number of available user in that environment. A scheduling algorithm accompanied by artificial neural network can be designed for setting the appropriate power level given as output from the neural network trained with data obtained over a period of time. Artificial neural network (ANN) computations unlike von Neumann model do not require separate memory or processing as it operates via the flow of signals through the net connections termed neurons, somewhat akin to biological networks. This artificial neural network is used in predictive modeling, adaptive control, and applications where sufficient datasets are available to train the neurons.Unidirectional Artificial feed forward neural without loops is the first simple ANN which propagates data from input to output via hidden nodes in hidden layer. Reliable wireless communication between the network and end device is achieved using BTS. The end devices like computer, mobile phone, and wireless networks like CDMA, Wi-Fi, and

GSM. BTS transmit waves that serve a purpose of enabling long distance coverage, but,at the same time pose a threat to the humankind because of its harmful EMF content. This paper deals with dynamically switching of the power levels of BTS according to the requirement. The network is trained to release a power level as an output based on the number of users in that environment. One out of three different power levels to be set for BTS, is estimated as an output from the trained neural network by giving derived parameters as input for analyzing the environment. The work also proposes a scheduling algorithm which takes the output of the ANN and gives the appropriate power level to be set at the corresponding time period.

2 Related Work

The scheduling of power levels in the telecommunication networks has been carried out in the various perspectives. One such proposal is made for the WiMax Medium Access Control (MAC) supporting wireless communication utilizes high bandwidth and its QoS requirements vary based on the type of applications employed [1]. MAC does not have a scheduling algorithm to achieve the objective of fairness, QoS and throughput. To satisfy multidimensional objective Scheduling Algorithm along with ANN and Fuzzy is used as it addresses the above mentioned objectives simultaneously. Fairness is achieved as a result, among the users by using the algorithm while keeping priority intact. In addition to this, high channel using up is attained with less computation time. Computer assisted Collaborative Work improvement in effectively combining is considered [2] and for which routers supporting multicast traffic facility is needed. Differentiation between unicast and multicast is not attained with queuing or scheduling. ANN is deployed in power plants for scheduling [3] the needed power to be generated for fulfilling the demands of consumer by prediction using the current load level automatically. Energy reliable wireless communication system has to be designed as increase in energy has led to increased CO_2 emission. The number of users belongs to a particular base station is low, then the base station is switched off and the users are served by neighboring stations. Through this concept 20% [4] of energy consumption is reduced. The drastic growth in telecommunication has resulted in increased health hazards due to high electromagnetic pollution. As a step towards Green Mobile Computing, proposes a model for electromagnetic pollution index and derives the factors influencing the index in order to manage the pollution. The formulation of electromagnetic pollution is made through a packet of pollution (PoP) which helps to devise the methodology to reduce EM pollution index in [5]. ANN using back propagation learning is designed to automate the accurate scheduling of power generation plant rating 4×8 MW [6] depending on the load.

3 Power Control at BTS Transceiver

Base Transceiver Station offers reliable wireless connectivity between telecom network and device of end users. Transceiver in BTS is equipped with power amplifier for amplifying the signals before relaying it to the antenna. The maximum transmit power at BTS is determined by the power amplifier rating in BTS. The load of the base station is determined by differences in maximum and present transmission power. The transmitting power level is adjusted based on number of users in the region sufficiently to attain required signal strength. Strong signal strength is needed for end user away from station than the one close to it. Suppose, for a given signal strength if quality of voice is more than sufficient, then the strength can be reduced. Similarly, if there are less number of users in the region than the power level can be reduced considerably. Thus the concept of power control at BTS gives rise to advantages like reduced power consumption, interference, and reduced radiation exposure period. While adjusting the transmission power at BTS the carrier signal strength to interference ratio must be considered for an acceptable power level. Base transceiver station gives 250nW of power near 1 GHZ and 1 μW of power near 12 GHZ [7]. The power generated by BTS is usually constant in a particular frequency band. The model proposed in the system adapts the transmitting power level of transceiver depending on the number of users in particular region of the day.Adapting the transceiver power level helps to reduce unnecessary radiations in the environment where there are only low number of users. The amount of energy in the emitted radiation is also considerably reduced so the ill effects due to power level can be decreased. The radiation exposure for a long time results in some drastic effects like blood pressure, miscarriage, depression, and DNA injury. The proposed model thus reduces the presence of radiations rich in energy by switching the power levels. The constant radiation exposure for the long term is thus prevented by varying the power levels of radiations in densely populated areas. The mobile users are considerably more in daytime when compared to night time, therefore the power level required during the day is significantly more than the power required to cover the users at night time. So the power level can be reduced during night time as it is sufficient to connect with few mobile users. This has resulted in reduced power consumption and shortened radiation exposure period. The conservation of power in turn will reduce the health hazards due to the effects of energy levels in harsh EMF radiations emitted by BTS placed in densely populated areas.

4 Neural Network Design

The artificial neural network has input where a number of users are given, hidden for processing and output layer which gives the output power level. The neural network [8] creates the network and train the network using input output values. It uses mean square error and regression models for analyzing the performance of the system. The neural network, thus maps the set of given inputs to the set of target values. Later, the target values can be derived for further processing. This work creates neural network [9, 10] to train with the input being number of users and the target being power levels assigned for different range of users. The network is initially trained with datasets having a different range of users as input and correspondingly the target output as power levels collected from local stations for about a week. Once the neural network is trained with analyzed datasets, it is simulated with custom inputs to cross check whether it is working properly. The weighted output of the neural network is the power level, which is to be assigned at BTS for a particular range of users.

The data about the number of users and power level required to satisfy their QoS requirements are collected over a week from local custom base stations. The collected data for every 24 hours is classified as day, afternoon, evening, and night time. The number of users and their power consumption in BTS are analyzed using the collected data. The number of users in day and evening will be more when compared to the number of users at night time. Traffic will also be higher during evening time. The collected dataset over a week about the number of users, time period, and respected power consumptions are given as input to the neural network. After training the network, its performance is measured in terms of mean square error and regression plot as shown in Figures 1 and 2. The power level increase in user and, when the number of user level drops the power level can be reduced which reduces signal strength in that environment. From the Figure 1, it is noticed that the regression value of the plot is very close to 1, which means the error is minimized to a great extent. The regression value is brought closer to 1 by training the Neural Network with a bigger Dataset. The more the datasets [11] given for training the network, the more accurate the results will be. So the number of possible combinations of verified input and target is given to train the network.

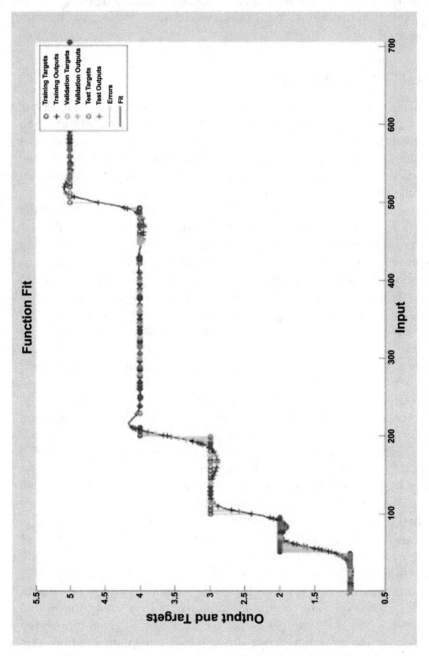

Figure 1   Number of users *vs* target power level.

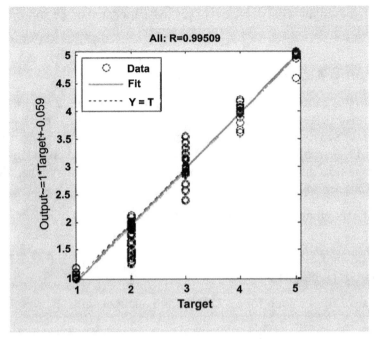

Figure 2 Regression plot.

The dataset given as input for training the neural network is divided into three datasets, as training, validation and testing. The network initially uses training datasets to create a neural mathematical model for the specified input and target.

After creating the model using training data, the neural network uses validation datasets to validate and correct at each step based on the evaluated results. Finally, the test vectors [12] are used to determine the performance of the modeled neural network by evaluating the number of errors. The mean square error can be considerably reduced when more datasets are given for training and the number of iterations required to converge also plays a major role. The Figure 1 shows three different vectors and their correlation with trained values. Figure 2 shows power level assigned to a specific range of user in a town. For e.g. for a range of 1–49 number of users, the power level assigned is 1. Likewise, there are power levels 2, 3, 4, 5. As the number of users increases the power level is accordingly increased for a particular time of the day. The neural network is now capable of giving power level based on any number of user given to it as input. This power level has to be allocated in the appropriate time slot to the BTS.

5 Scheduling Algorithm for Power Switching

The scheduling algorithm is basically used to switch the power levels in BTS. The algorithm allocates identified power level in corresponding time slots and switches the BTS to low, medium, or high based on the input provided to ANN. The priority levels are set during day time and increased based on the mobile users in the region due to local functions. The following is the scheduling algorithm designed to allocate the power level.

Step 1: Get the number of users in the cell periodically as input based on some QoS parameter.

Step 2: Create an if-else branch with conditions created by 3 time periods as day, evening, or night. (A)

Step 3: Create a similar loop for packet size and queue length. (B)

Step 4: From the priority levels received in (A) and (B), we create a priority index using if-else branch with three levels namely L, M, and H

Step 5: The output received is the final priority index showing one among the three available power levels

The priority index points to the power level and packets to be sent to the queue. The appropriate packet to be sent based on the priority index determined using A and B. Thus the power level is determined based on number of users in the region and BTS is switched to the respective power level for a period of time to transmit the packets.

6 Implementation

Along with the scheduling algorithm, this paper also involves an extra parameter known as ILONP (Importance Level of Network Provider). The function of this variable is to combine the functions of all the network providers to one BTS. This variable will decide the priority levels of different network providers. Based on their ILONP values, their request for the frequency band will be met with. This variable combines with the final priority level and gives the decision parameter to the BTS as to whether it should serve the request or not. Once the Neural Network is trained and simulated, the model is saved as a network. This model releases an output file whenever it is fed with an input. The error rate for a neural network cannot be more than 0.1. And with this fact, we can easily assume that once we round off the value of power level received from the historical database, it will be the correct value for sure. This power level is taken as Parameter 1. Along with this, three other parameters are left to the network provider to choose from. This will help the Network provider to bring in QoS (Quality of Service) factors at the scheduling level.

Excel file, it will be the correct value for sure. This power level is taken as Parameter 1. Along with this, three other parameters are left to the network provider to choose from. This will help the Network provider to bring in QoS (Quality of Service) factors at the scheduling level.

7 Results and Discussion

For the analysis of the prototype, the models were tested with data from the industry. The regression value came close to 0.99 which shows that the model will give near perfect values. The Java code also functioned without any glitches. The prototype, when fully functional, will surely save a lot of power and hopefully save some lives by preventing harmful EMF radiations to spread. This work models, neural network in order to predict the required sum of power level based on the number of users at any given point of time. Back propagation [13] based learning technique gives good result and the amount of data given for training plays an important role. If a sufficient quantity of data is offered to learn, then the network will provide accurate results. The number of hidden nodes impacts the difference in the error between target and estimated value. Also, in our case the scheduling algorithm provides a rounding off function which eliminates the error completely and provides a base for further expansion by providing three parameters for the network providers to decide upon.

8 Conclusion

This work constructs soft computing based scheduling technique to resolve any uncertainties present in the outputs which may evolve as errors in the model and interfere with the actual purpose for which the application is deployed. Using ANN to accomplish dynamic switching of power levels, the level of uncertainty is computed and number of inspections were made to identify the correlation of output with the given input values. The proposed scheduling algorithm along with neural network reduces the exposure time of the constant high level radiations by switching to low powers where the users are low in number. Hence, the signal strength in the environment and the impact of radiations on living beings can be considerably reduced. The sufficient signal amount provided for transmission also reduces the effect of interference. Therefore the proposed algorithm confirms to green mobile computing by reducing the strength of radiation in the environment through adapting the power level.

References

[1] D. David Neels Pon Kumar, K. Murugesan, S. Raghavan, and M. Suganthi, "Neural Network based Scheduling Algorithm for WiMAX with improved QoS Constraints", International Conference on Emerging Trends in Electrical and Computer Technology (ICETECT), 2011.

[2] Malika Bourenane, "Energy-Efficient Scheduling Scheme Using Reinforcement Learning in Wireless Ad Hoc Networks", Eighth International Conference on Wireless and Optical Communications Networks (WOCN), 2011.

[3] Meina Song, Xiaosu Zhan and Junde Song "An Efficient Queueing Scheme for Multicast Packet Switching Routers", Eigth International conference on Computer Supported Cooperative Work in Design, 2004.

[4] Prithiviraj Venkatapathy, J Jena, Avadhanulu Jandhyala, "Electromagnetic Pollution Index–A Key Attribute of Green Mobile Communications", Green Technologies Conference (GTC), IEEE, pp. 1–4, 19–20 April 2012.

[5] Prithiviraj, V., Venkatraman, S. B., and Vijayasarathi, R, "Cell zooming for energy efficient wireless cellular network", Journal of Green Engineering, Vol 3, (4), pp. 421–434, 2013.

[6] Mahmoud Moghavvemi, S. S. Yang, and M. A. Kashem, "A Practical Neural Network Approach For Power Generation Automation", Proceedings of EMPD '98 International Conference on Energy Management and Power Delivery, Vol. 1, 1998.

[7] GSM Technical specification, European Telecommunications Standards Institute 1996.

[8] B.Yegnanarayana, Artificial Neural Networks, PHI Learning Pvt. Ltd, 2004.

[9] Satish Kumar, "Neural Networks: A Classroom Approach", Tata McGraw-Hill, ISBN: 978-0-07048292-02004, 2004.

[10] James. A. Freeman, "Neural Networks: Algorithms, Applications, and Programming Techniques", Pearson Education India, 1991.

[11] S. Geman, E. Bienenstock, and T. Doursat, "Neural networks and the bias/variance dilemma," Neural Comput., vol. 5, pp. 1–58, 1992.

[12] H. Gish, "A probabilistic approach to the understanding and training of neural network classifiers," in Proc. IEEE Int. Conf. Acoustic, Speech,Signal Processing, 1990, pp. 1361–1364.

[13] L. W. Glorfeld, "A methodology for simplification and interpretation of back propagation-based neural networks models," Expert Syst. Applicat., vol. 10, pp. 37–54, 1996.

Biographies

Dr. M. Hemalatha is graduated in Electronics and Communication Engineering, post graduated in Information Technology and Ph.D in Broad band Wireless Communication in the year 1997,2003 and 2012 respectively. She had been associated with SASTRA University, Thanjavur, TN, India, for about 15 years and currently working as Professor in the department of ECE, Narayana Engineering College, Nellore, AP, and India. Having 17+ years of teaching experience, has delivered various technical workshops, lectures and conducted aided projects. Her keen interest in wireless communication and baseband signal processing driven to publish more than 30 articles in the national and international conferences and international journals.

Dr. V. Prithiviraj received M.S degree in Electrical Engineering from IIT Madras., Ph.D. in Electronics and Electrical Communication Engineering from IIT Kharagpur. He is working as Principal Rajalakshmi Institute of Technology from May 2013. He has over 3 decades of teaching experience and 12 years of Research & Development Experience between the two IITs in the field of RF & Microwave Engineering. His areas of interest include Broadband and Wireless Communication, Telemedicine, e-Governance and Internet of Things.

Dr. Jayasri. R has received B.E(ECE) from Anna University, TN, India and M.Tech(Embedded Systems) from SASTRA University. Having good number of publications in her earlier career, is now working as assistant system engineer in Tata Consultancy Services, Chennai, India.

www.ingramcontent.com/pod-product-compliance
Lightning Source LLC
LaVergne TN
LVHW012333060326
832902LV00011B/1866